AF184042

Advances in Spectroscopic Analysis of Food and Drink

Online at: https://doi.org/10.1088/978-0-7503-5573-5

IOP Series in Spectroscopic Methods and Applications

Series Editor
Ashutosh Kumar Shukla, Ewing Christian College, University of Allahabad, Prayagraj, India
Daniel Cozzolino, The University of Queensland, Brisbane, Australia

About the Series

The *IOP Series in Spectroscopic Methods and Applications* aims to develop a library of key texts encompassing the broad range of spectroscopic techniques, instrumentation, and fields of applications found in modern science. Some examples of the spectroscopic techniques incorporated include, IR, UV/VIS, Raman, EPR, NMR, fluorescence spectroscopy, force spectroscopy, hyperspectral imaging, neutron scattering, Mössbauer spectroscopy, photoemission spectroscopy, and all the various specialist techniques within these broad domains and more. Microscopic characterization techniques may also be included where they are used in conjunction with spectroscopic methods for sample analysis.

The series contains two broad types of approach: those addressing a particular field of application and review the numerous relevant spectroscopic methods applicable to the field, and those that focus on a specific spectroscopic method that will permit a greater in-depth review of the theory and instrumentation technology.

A full list of titles published in this series can be found here: https://iopscience.iop.org/bookListInfo/iop-series-in-spectroscopic-methods-and-applications.

Advances in Spectroscopic Analysis of Food and Drink

Edited by
Ashutosh Kumar Shukla
Department of Physics, Ewing Christian College, Prayagraj, UP, India

IOP Publishing, Bristol, UK

© IOP Publishing Ltd 2023

All rights reserved. No part of this publication may be reproduced, stored in a retrieval system or transmitted in any form or by any means, electronic, mechanical, photocopying, recording or otherwise, without the prior permission of the publisher, or as expressly permitted by law or under terms agreed with the appropriate rights organization. Multiple copying is permitted in accordance with the terms of licences issued by the Copyright Licensing Agency, the Copyright Clearance Centre and other reproduction rights organizations.

Permission to make use of IOP Publishing content other than as set out above may be sought at permissions@ioppublishing.org.

Ashutosh Kumar Shukla has asserted his right to be identified as the editor of this work in accordance with sections 77 and 78 of the Copyright, Designs and Patents Act 1988.

ISBN 978-0-7503-5573-5 (ebook)
ISBN 978-0-7503-5571-1 (print)
ISBN 978-0-7503-5574-2 (myPrint)
ISBN 978-0-7503-5572-8 (mobi)

DOI 10.1088/978-0-7503-5573-5

Version: 20240201

IOP ebooks

British Library Cataloguing-in-Publication Data: A catalogue record for this book is available from the British Library.

Published by IOP Publishing, wholly owned by The Institute of Physics, London

IOP Publishing, No.2 The Distillery, Glassfields, Avon Street, Bristol, BS2 0GR, UK

US Office: IOP Publishing, Inc., 190 North Independence Mall West, Suite 601, Philadelphia, PA 19106, USA

To my parents

Contents

Preface

Quality analysis of food and drink is undoubtedly in everyone's interest. Accordingly, this field has consistently attracted the attention of the scientific community. The quality parameters of these items may be analysed using several approaches. This volume, titled Advances in Spectroscopic Analysis of Food and Drink, presents quality assessment applications based on different spectroscopic techniques. Fruits, fruit juices, coffee beans, and edible oil samples have been covered in particular. Case studies on grape fruit juice, pomegranate fruit juice, and apple juice have been presented among the fruit juice category. Similarly, quality analysis of varieties of coffee (espresso, traditional, and instant), mate tea, star anise, and their respective infusions have been presented. There are 12 chapters in this volume covering different molecular and atomic spectroscopy techniques, including Fourier transform infrared (FTIR) spectroscopy, near-infrared (NIR) and mid-infrared (MIR) spectroscopy techniques. Electron spin resonance (ESR) spectroscopy techniques have also been included. The chapters have mostly adopted a bottom-up approach and talk about the applications of these spectroscopic techniques combined with data analytics.

I sincerely thank John Navas, Senior Commissioning Manager at IOP Publishing, for giving me an opportunity to present this volume. I also wish to thank Phoebe Hooper, Editorial Assistant at IOP Publishing, for extending her support during different stages of this project. I thank the expert contributors for their quality contributions, which have made this volume a unique collection.

Ashutosh Kumar Shukla
Prayagraj, India

Editor biography

Ashutosh Kumar Shukla

Ashutosh Kumar Shukla has more than two decades of physics teaching and research experience. He has published numerous articles and review articles in peer-reviewed journals. He has also authored textbooks and participated as an editor of more than two dozen edited volumes published by reputed publishers, often prepared in collaboration with experts from different countries. He intends to continue pursuing his academic interests, which include spectroscopy applications in different fields of societal importance.

List of contributors

Roney Alves da Rocha
Engineerng Department, Federal University of Lavras, Lavras, Brazil

Grasieli Beloni de Melo
Faculty of Food Engineering, Department of Food Science, University of Campinas, São Paulo, Brazil

Nor Kartini Binti Abu Bakar
Department of Chemistry, Universiti Malaya, Kuala Lumpur, Malaysia

Daniel Cozzolino
The University of Queensland, Centre for Nutrition and Food Sciences, Queensland Alliance for Agriculture and Food Innovation (QAAFI), Brisbane, Australia

B Dayananda
The University of Queensland, Centre for Nutrition and Food Sciences, Queensland Alliance for Agriculture and Food Innovation (QAAFI), Brisbane, Australia

Fabiana de Carvalho Pires
Engineerng Department, Federal University of Lavras, Lavras, Brazil

Sumaiya Fatima
Department of Food Technology, School of Chemical Technology, Harcourt Butler Technical University, Kanpur, India

Rosemary Gualberto Fonseca Alvarenga Pereira
Engineerng Department, Federal University of Lavras, Lavras, Brazil

Grzegorz Guzik
Institute of Nuclear Chemistry and Technology, Warsaw, Poland

Nabamita Halder
Department of Life Science, NIT Rourkela, Rourkela, India

Agustina A M B Hastuti
Department of Pharmaceutical Chemistry, Universitas Gadjah Mada, Yogyakarta, Indonesia

Gunawan Indrayanto
Faculty of Pharmacy, Airlangga University, Surabaya, Indonesia

Juliana Juliana Azevedo Lima Pallone
Department of Food Science, University of Campinas, São Paulo, Brazil

Rasool Khodabakhshian
Department of Biosystems Engineering, Ferdowsi University of Mashhad, Mashhad, Iran

Vivek Kumar
Department of Food Technology, School of Chemical Technology, Harcourt Butler Technical University, Kanpur, India

Monalisa Mishra
Department of Life Science, NIT Rourkela, Rourkela, India

Abdul Rahman
G N Ramachandran Protein Centre, CSIR- Institute of Microbial Technology, Chandigarh, India

Michel Rocha Baqueta
Universidade Tecnológica Federal do Paraná, Paraná, Brazil

Abdul Rohman
Faculty of Pharmacy, Gadjah Mada University, Yogyakarta, Indonesia

Triveni Shelke
Department of Life Science, NIT Rourkela, Rourkela, India

Wacław Stachowicz
Scientist emeritus, Institute of Nuclear Chemistry and Technology, Warsaw, Poland

Elem Tamirys dos Santos Caramês
Department of Food Science, University of Campinas, São Paulo, Brazil

Deepika Umrao
Department of Biotechnology, Dr B R Ambedkar National Institute of Technology Jalandhar, Jalandhar, India

Patrícia Valderrama
Universidade Tecnológica Federal do Paraná, Paraná, Brazil

Anjar Windarsih
Department of Chemistry, University Malaya, Kuala Lumpur, Malaysia and Research Center for Food Technology and Processing (PRTPP), National Research and Innovation Agency (BRIN), Yogyakarta, Indonesia

Prinya Wongsa
School of Agro-Industry, Mae Fah Luang University, Thailand

Lestyo Wulandari
Faculty of Pharmacy, Airlangga University, Surabaya, and Faculty of Pharmacy, State University of Jember, Jember, Indonesia

Mohammad Yuwono
Faculty of Pharmacy, Airlangga University, Surabaya, Indonesia

Chapter 1

Determination of fruit quality: a spectroscopic approach

Triveni Shelke, Nabamita Halder and Monalisa Mishra

The market price of fruit is significantly influenced by quality. Meanwhile, the quality of fruit may be improved by a variety of processes that concentrate on firmness, ripening, acidity, and soluble solid contents, among other factors. However, the different procedures that claim to improve the quality of fruit need to be examined. The parameters that are used in determining the quality of fruit can be assessed using spectroscopic analysis. Since existing techniques are labor-intensive and complicated, spectroscopy offers a superior way to ascertain the chemical and physical characteristics of fruit. In particular, the internal and external chemo-matrices of the fruit can be identified using the spectroscopic method. Adulterations of fruit products can also be an analyzed using spectroscopic methods. Fruits such as mango hold a strong place in the Indian market and spectroscopic approaches can help to improve the yield of better-quality fruits.

1.1 Importance of fruit in the human diet

Fruits can be classified based on their anatomical structures and origins as simple, complex, and accessory fruits. The simple fruits include dry and fleshy simple fruits [1]. On the basis of the region in which they grow, they are classified as temperate zone, sub-tropical and tropical fruits (figure 1.1). Fruits are nutritionally rich (table 1.1) and are a source of dietary fibers, vitamins, and minerals [2]. They are strong antioxidants, and act as detoxifiers of carcinogens and modifiers of metabolic processes. These properties make fruits beneficial supplements to avoid diseases such as heart disorders, chronic diseases, and cancer [3]. The dietary fibers form bulks in the intestine, and hence increase absorption of nutrients. The fibers when fermented to short fatty acids in the colon act as anti-carcinogens [4]. In addition, high fiber content helps in calcium absorption and decreases acid-load in the diet [5]. Several secondary metabolites, e.g., anthocyanins, procyanidins, and flavanols, reduce

doi:10.1088/978-0-7503-5573-5ch1 © IOP Publishing Ltd 2023

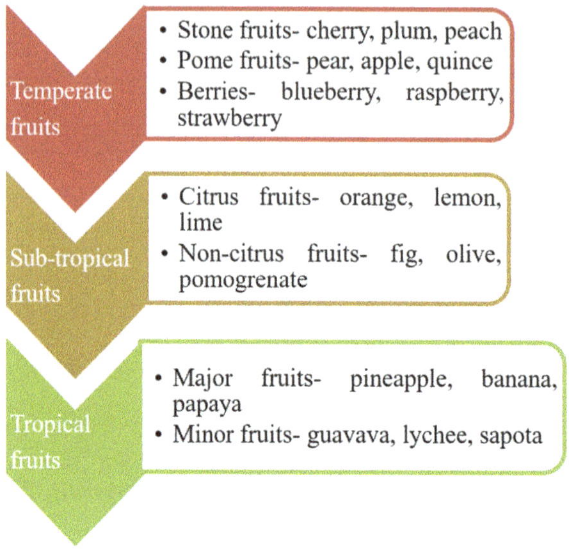

Figure 1.1. Classification of fruits based on their regions of growth.

Table 1.1. Nutritional qualities of fruits.

Nutrients	Examples	References
Carbohydrates	Fructose, glucose	Slavin, Lloyd [8]
Antioxidants	Carotenoids, phenolic compounds	
Dietary fibers	Cellulose, pectin, lignin	
Minerals	Calcium, iron, potassium	
Vitamins	Vit A, B6, C	
Others	Water, proteins, fats	

heart-related disease risks, and protect against lens damage and respiratory syndromes [6]. Carotenoid rich fruits maintain cholesterol and reduce oxidative damage [7].

Low consumption of fruits is associated with diseases such as hypercholesterolemia, blood pressure, osteoporosis, and pulmonary diseases [9], it also promotes obesity due to overgrowth of adipose tissues [10]. Fruits help in loss of body weight because the fibers create an environment in which the enzymes hydrolyzing fats, carbohydrates, and proteins have reduced activity [11]. The deteriorating quality of fruits is one of the reasons for increasing malnutrition [12]. The recommended level of fruit and vegetable intake is 400 g/person [13]. The 2015 data of GEMS/Food database shows the mean fruit intake of India was 158.2 g/d [14]. The quality of fruit deteriorates due to many factors that influence the production, ripening, and marketing of the fruits.

1.2 Factors affecting fruit quality

The quality of fruit is determined by various factors and is tested on several parameters (figure 1.2).

1.2.1 Pre-harvest factors

Pre-harvest factors include environmental and cultural factors.

1.2.1.1 Environmental factors

Temperature determines the rate of transpiration, and effects photosynthesis, uptake of nutrients, biochemical reactions inside the plant cells, and fruit setting [15]. Temperature affects the ripening of the fruits: some fruits when exposed to sunlight ripen faster, whereas others delay their ripening. Temperature also influences the sugar content and acidity [16]. The concentration of carbon dioxide is related to the stomata conduction and ozone uptake [17]. Optimum rainfall is essential for better-quality fruit: higher rainfall can cause mechanical damage to the fruits, whereas lesser rainfall can affect the growth of the plant [18]. Exposure to sunlight affects the size, weight, and anthocyanin production in fruits. In addition, the canopy light condition influences the time of ripening, and hence maturity [19]. Finally, humidity is related to the color, acidity, and also Salmonella outbreaks in fruits [20].

1.2.1.2 Cultural factors

The cultural factors take nutrition, growth regulators, and disease-causing pests into consideration. The appropriate amount of minerals such as nitrogen, calcium, phosphorus, boron, copper, magnesium, and potassium are essential for fruit development. The other compounds that are actively involved in the growth of the fruits are the growth regulators or phytohormones, which constitute auxin, cytokinins, ethylene, and gibberlins [21]. Pests such as Saltatoria, Dermaptera, Isoptera, and Hemiptera are widely responsible for fruit stock losses [22].

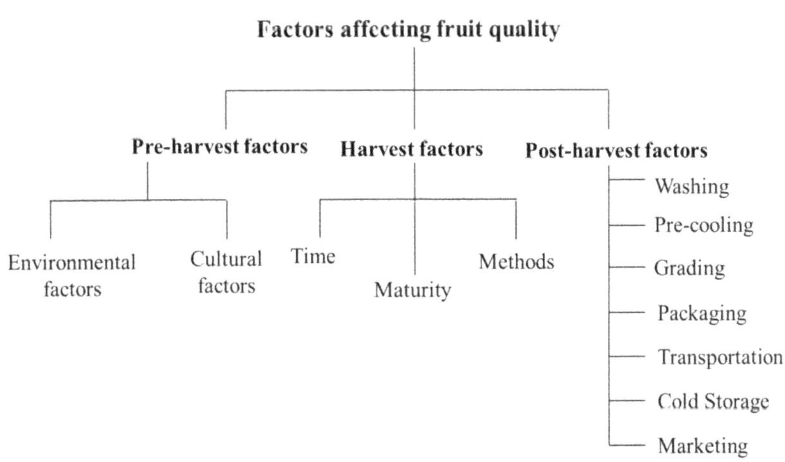

Figure 1.2. Factors affecting fruit quality.

1.2.2 Harvest factors

The harvest factors include the time and maturity at which the fruits are harvested, as well as the methods used for harvesting. The time of the day for harvesting, and the time lapse between harvest and transportation make the fruit susceptible to bruising. The damage done by bruising is more prominent when the harvesting is done at the heat prone time of the day. Bruising decreases the shelf life of the fruit because it promotes early senescence [23]. Meanwhile, an early or late harvest influences the fatty acid composition in the fruits [24]. Fruits are divided on the basis of maturity as fully mature, half mature, and immature. At the time of harvest, the fully and half mature fruits show good fruit quality upon ripening when compared to the immature fruits [25]. Berries are generally harvested at early maturity stages for their firmness [26]. The traditional methods of harvesting with hooks attached to bamboo may damage the fruit due to injuries and bruises [27]. Sun drying of the berries upon using impact force harvesting leads to a fading of fruit color and affects its appearance; therefore, the shade drying method is much preferred for berries [28]. Modified machine harvesters decrease the firmness and increase the probability of bruises in the berries [29].

1.2.3 Post-harvest factors

Post-harvest factors include packing house operations such as washing, pre-cooling, grading, packaging, transportation, cold storage, and marketing. The fruits are washed with chlorine water around the pH of 6.5–7.5 to maintain quality. Pre-cooling removes the heat acquired by the fruit when in the field, which enhances shelf life. Various types of pre-cooling methods are applied to the fruits based on their requirements. The grading of the fruits is based on their size and appearance. Instruments such as a size gauge and rings for circular fruits are used to grade them into a designated category. The type of packaging and the arrangement of the fruits in layers from top to bottom determine the fate of the fruit in the market. Generally, the fruits in the top layers do not suffer much damage whereas those in the bottom layers suffer bruises [30]. Packing materials such as double walled fibers, foam nets, and plastic trays to arrange the fruits in layers are chosen according to the texture of the fruit [31]. Temperature has to be monitored during the transportation of fruits to the storage houses, and eventually to the market. In addition, radiation treatment and fumigation are usually used to resist certain molds that can infect the fruits during transportation [32]. Cold storage can alter the fruit's softness, antioxidants, and biologically active compound levels depending on the pre-applicants used [33]. There are also many different stages of marketing that can affect the fruit's quality [34].

1.3 Spectroscopy in fruit quality determination

Fruit quality plays an essential role in customer satisfaction and fruit sales as a whole. Fruit quality is mainly attributed to ripeness, color, taste, smell, firmness, soluble solid contents (SSC), and size. These are common physical markers that are used to assess the condition of a fruit before purchase but they are greatly unreliable and do not give an absolute assessment of the fruit's quality, hence more competent parameters have been taken into consideration, such as acidity, sugar concentration, and starch content. The SSC, sugar concentration, acidity, and starch content measurements are estimated by conventional means such as refractometry, pH monitoring, and titration acidity [35]. However, these procedures are invasive, difficult to perform, and expensive [36]. This has prompted the demand for a reliable, non-invasive, and fast method for fruit quality assessment.

The most effective method for accomplishing this is spectroscopic analysis, which has many benefits: it is non-invasive, which allows for the acquisition of internal quality parameters in fruits without affecting their surfaces; the quantification processes are straightforward and fast because no time-consuming chemical treatments or reactions on fruit samples are needed; and it enables the simultaneous detection of multiple internal fruit attributes [37]. However, the disadvantage of the tiny point-source measurements frequently used in spectral assessment is that they do not provide spatial information, which is essential in many scenarios evaluating fruit quality [37]. To accurately quantify the internal parameters of fruit quality, the visual (Vis) and near-infrared (NIR) properties are typically used in the analysis of fruit quality. Hyperspectral imaging, a combined spectroscopic and imaging approach that produces a three-dimensional hyperspectral image with two-dimensional spatial information and one-dimensional spectral information, is another technique that is used to assess the quality of fruit [38]. The commonly used techniques for fruit quality analysis are explained in the following subsections.

1.3.1 Vis/NIR spectroscopy

Vis/NIRS spectroscopy has a lot of applications in determining fruit quality. It has been used to acquire fruit spectra from both climacteric fruits such as mango, apple, banana, plum, apricot, peach, nectarine, and kiwi, and non-climacteric fruits such as watermelon, mandarin, and cherry. The NIR region of the electromagnetic spectrum has a range of 750–2500 nm and measures internal parameters such as titrable acidity (TA), ripeness, SSC, pH, and other fruit quality parameters in a non-destructive way. The visible region of the electromagnetic spectrum has a range of 400–750 nm, and measures the color and pigments present in the fruit. Several studies of the assessment of these properties have been conducted, which have assessed attributes at a certain wavelength. For example, a portable Vis/NIR spectrometer with a range of 960–1700 nm has been used to determine the SSC and TA 'Tommy Atkins' mango [39]. This works on the principle that the Vis/NIR has the signals for almost all of the functional groups and structures of the compounds with a probable stable spectrum [40], and the varying wavelengths are compatible with specific compounds, such as organic compounds with –OH, –NH, and –CH, bonds, are more sensitive to wavelengths in the NIR range [41]. Wavebands found in the Vis/NIR range are frequently used in multispectral and hyperspectral imaging techniques to assess fruit quality [42–44]. The detection of pigments such as chlorophyll in immature canola seeds [45] and carotenoids in wheat [46] is made possible by spectrophotometric examination of absorbance at certain wavelengths associated with the pigments. Radiation that strikes a specimen can be reflected, transmitted, or absorbed, which generates a specific spectrum reflecting a physical property or chemical constitution of the material [37]. This makes it easier to sequester the comparable parameters for testing and quantifying. The Vis/NIR range includes wavelengths for most of the major functional groups, side chains, and chemical compounds. This region is also highly dominated by H_2O, which absorbs in the NIR region and leads to interference in the spectral data. This is further followed by chemometric techniques, which involve the quantification of the desired variable and removal of unrelated variables from the spectral data collected from the sample. Chemometrics aid in obtaining clearer information from the spectral data regarding the required characteristics of the fruit samples.

The Vis/NIR range also has the advantage of quantifying more than one parameter at the same time, hence replacing several tests while also reducing contact procedures and human effort. This method appears to be effective in assisting breeders with genotype screening in breeding operations. Additionally, it enables online fruit screening and fruit quality evaluation, creating new market niches and prospects for fruit commercialization in the raw or processed market [47]. Now that Vis/NIR technology is available in handheld devices, it is becoming increasingly common to apply Vis/NIR in fruit quality analysis techniques.

1.3.2 Hyperspectral imaging spectroscopy

Classification systems for fruits and vegetables use a color video camera to take photographs with three filters that mimic human vision by employing red, green, and blue (RGB) wavelengths [48, 49]. They are therefore limited to observing events and are frequently unable to learn much about the exterior or inside makeup of the items. They are also unable to distinguish outward damage that has a color comparable to the healthy fruit skin. As a result, the market's increasing demand for fruits of higher quality has contributed significantly to the development of methods that can extract both spectral and spatial data from a sample using spectroscopy and imaging techniques such as hyperspectral imaging (HSI) spectroscopy [50]. These techniques are similar to the Vis/NIR in terms of quick evaluation and non-invasive determination but they have distinct differences with regards to their spectral data output—Vis/NIR techniques obtain their spectral data from a point-source, which limits the scanning to a small area of the sample, whereas HSI scans over a larger area, which gives more inherent knowledge about the components of the sample and works with its heterogeneity [51].

Since it is a result of the physical and chemical characteristics of the individual substance examined, the spectral signature obtained through HSI is distinctive [52]. Hyperspectral images are three-dimensional cubes made up of one wavelength and two spatial dimensions [$I(\lambda, x, y)$] [53]. Four methods can be used to create three-dimensional hyperspectral image cubes. There are two spectral scanning methods and two spatial scanning methods: area and plane scanning, point scanning, and line scanning, respectively [54]. When compared to VIS/IR analysis, which examines the sample as a whole and calculates its average composition, HSI has the advantage of providing spatial dispersion of sample quality metrics [55]. HSI can detect several injuries. It can also detect decay in the fruit and vegetable samples with great accuracy, e.g., bruising of fruits is not as readily detected by other spectroscopic methods due to variance in the defects and lack of proper spectral data. For example, Xing *et al* tested the effectiveness of a hyperspectral imaging system for detecting apple bruises [56]. The HSI proves to be more accurate over NIRS because it provides a continuous evaluation in production with greater data of the variety of attributes of fruits. HSI calibrations are also faster due to their greater statistical data from a single sample, where NIRS assumes sample homogeneity which might not be the real case [57].

The only drawback of HSI is that its setup is quite costly and time consuming. In addition, the hyperspectral imaging system cannot be employed online or in real-time due to the redundant data. In addition, a multispectral imaging system's effective wavelengths are always determined by analyzing the data from hyperspectral imaging [58].

1.4 Chemometrics analyzed through spectroscopy

The data that are usually obtained from various spectroscopy techniques are mostly obtained from the Vis/NIR region. This region (as discussed earlier) includes the wavelengths for almost all of the side chains. However, the presence of so much information leads to undesired complications in the spectrum. Hence, it is important to preprocess the spectral data to improve the results and reduce noise, so that it can be analyzed properly [59]. There are many techniques for spectrum preprocessing and calibration in chemometrics, including partial least squares (PLS), principal component regression (PCR), and others. Note that MLR represents multiple linear regression, LS-SVM represents least square support vector machine, and ANN represents artificial neural network [41, 60]. The different types of pre-treatment methods are discussed in table 1.2.

1.5 Parameters for fruit quality analysis

Fruit quality may be measured by analyzing various parameters through spectroscopy (table 1.3). These parameters are applied to the types of fruits [37].

1.6 Determination of fruit quality

1.6.1 Determining the quality of mango

Mango (*Magnifera indica* L.) is one of the most widely popular climacteric fruits, and has a huge commercial impact on both Indian and foreign markets. The quality of mangoes is described mainly by their sweet and sour notes, and the ripe color of the pulp. Mangoes are best collected when they are mature-green and pre-climacteric [75]. The fruits are picked in two stages: the pre-harvest stage and the post-harvest stage. The post-harvest stage is when the ripening is complete, the quality of fruits differs on the basis of the picking time. The fruits that have reached

Table 1.2. Chemometrics in spectral data.

Spectral preprocessing methods	Function	References
Smoothing	Removes high-frequency disturbances from the spectrum and improves the ratio of signal to noise. The primary idea behind it is to 'average' or 'fit' a number of points inside a window in order to arrive at an ideal estimation value. The spectral resolution would be reduced the wider the window is. Therefore, picking the right window width is essential.	Wang *et al* [37]
Offset Correction	In this type of central processing, the average value of the first few wavelength points is taken out of each spectrum, e.?g., five. The baseline drift is the only thing that is modified by offset correction; the spectrum structure is left unaltered. Its major purpose is to lessen the impact of optical distance, instrument noise, and the detecting environment.	Wang *et al* [37]
De-trending	De-trending is a strategy to get rid of the spectrum's baseline drift, which is typically used along with traditional normal variate correction (SNV).	Wang *et al* [37]
Multiplicative scatter correction (MSC)	MSC is a modification technique that is used to account for additive or multiplicative effects in spectral data.	Barnes *et al* [61, 62])
Standard normal variate (SNV)	SNV is a row-oriented modification that may be able to eliminate multiplicative interferences from spectral data that are caused by scatter and particle size effects. SNV reduces the effects of dispersion by focusing and adjusting each individual spectrum.	(R. Barnes *et al* [63]; Barnes *et al* [61])
Spectral derivatives	Drifting and scattering are removed using the first and second derivatives, respectively. They can improve spectral resolution and sensitivity, eliminate background interference, and separate superposed peaks. Two frequently used methods for spectral derivatives are direct finite difference and Savitzky–Golay (S?-?G) derivatives.	Wang *et al* [37]
Net analyte preprocessing (NAP)	NAP is mainly used to extract spectral data of a specific component in a mixed spectrum.	Wang *et al* [37]

Wavelet transformation	The basic idea of wavelet transformation it is to combine sinusoidal waves with numerous different amplitudes, frequencies, and orientations from the original spectrum. The chemical signals from the sample are broken down into measurable components according to their frequencies by using a basic function.	Wang *et al* [37]

Optimal wavelength selection	Function	References
Successive projections algorithm (SPA)	SPA is a forward selection method that seeks out a wavelength with the least amount of duplicated information and collinearity by using simple operations in a vector space. SPA begins with one wavelength and keeps adding more until a certain number is reached. The bulk of the sample's spectral data could be accurately captured using the SPA-extracted wavelengths with the least amount of information overlap.	Araújo *et al* [37, 64, 65]
Regression coefficient (RC)	The RC value is indicative of how predictive the model's performance is. The absolute value of RC represents the characteristic wavelength.	Wang *et al* [37]
Loading weights (LW)	In LW, a particular wavelength is chosen, and these wavelengths are equal to the latent variables. The loading weight is obtained from the latent variables, which is predictive of the impact of wavelengths on the model.	Wang *et al* [37]
Genetic algorithm (GA)	GA is an algorithm that uses genetic operations based on a randomly selected generation, such as genetic selection, crossover, and mutations to obtain best outcomes. Through this, useless spectral data is removed and noise is significantly reduced.	Wang *et al* [37]
Competitive adaptive reweighted sampling (CARS)	CARS is a cutting-edge algorithm that chooses the best wavelength based primarily on Charles Darwin's concept of the 'survival of the fittest.' The adaptive reweighted sampling method of a PLS model is used to sequentially identify wavelengths with large absolute coefficients as CWs. Cross validation is used to create a number of variable subsets, and the one with the lowest root mean square error of cross validation (RMSECV) is chosen.	(H. Li *et al* [66]; Wang *et al* [37])

(*Continued*)

Table 1.2. (*Continued*)

Optimal wavelength selection	Function	References
Uninformative variable elimination (UVE)	UVE is a method for selecting variables based on an analysis of the PLS regression coefficient. Using calibration samples that are randomly selected, the technique first builds a large number of models and then evaluates each variable using the stability of the corresponding coefficients in these models. Variables with poor stability are eliminated.	Centner *et al* [37, 67, 68]
Calibration models	**Functions**	**References**
Multiple Linear Regressions (MLR)	MLR predicts the dependent variables as a linear combination of the spectral values at each wavelength point. The error between the predicted and measured values is diminished in a least squares sense. The effectiveness of MLR algorithms is reduced by multicollinearity between the variables in spectral analysis.	Wang *et al* [37]
Principal component regression (PCR)	In PCR, a small number of principal components (PCs) are found using principal component analysis (PCA). These PCs are used as predictors and to create an MLR model rather than using the original spectral data.	Wang *et al* [37]
Partial least squares regression (PLS)	PLS forecasts the dependent variables by sifting through the variables to find the smallest collection of orthogonal elements with the best predictive power. The importance for forecasting the dependent variables was used to order these orthogonal elements, also known as latent variables (LVs). PLS regression, which combines the ideas of MLR and PCA, is especially useful when there is multicollinearity between the variables and there are frequently fewer latent variables than in PCR regression.	Wang *et al* [37]

the ripening stage are of better quality than those that are preharvested [75]. However, the ripened fruits lack the lifespan required for long delivery distances. Thus, determining fruit quality is of utmost importance, and using spectroscopic methods is the best way to achieve it without destroying the fruit samples. Mangoes fit for export are usually rated based on their green stage [75]. However, the buyers

Table 1.3. Parameters used to assist the analysis of fruit quality.

Parameters	Uses	Spectral range	References
Acidity	Measures of organic acids.	5882–9900 nm	Bureau *et al* [69]
Soluble solids/ sugar contents (SSC)	Measures the total solid content in the given solution. In the case of fruits, it is mostly the sugar content.	800–1100 nm	Angra *et al* [70]
Firmness	Measures the crispness of the fruits.	500–1000 nm	Mendoza *et al* [43]
Bruise detection	The damage in the fruit's cell and tissues is noted.	380–1000 nm	Luo *et al* [71]
Total polyphenols	Measures polyphenols content in fruits.	400–2500 nm	Pissard *et al* [72]
Vitamin C content	Measures Vitamin C content in the fruits.	1333–1835 nm	Xia *et al* [73]
Pigments	Measures different families of pigments in fruits.	680 nm	Zude *et al* [74]

will eat mango fruits when they reach the maturity stage. In the interim, the employees will be able to assess the quality of the mango fruits solely based on their physical characteristics, such as color, firmness, size, and weight [76]. When the fruits are mature, such grading techniques at the green stage typically cannot ensure the quality of the flavors. Therefore, the categorization of quality is crucial.

Theanjumpol *et al* determined the quality of mangoes with the spectroscopic approach. The spectral data was collected at wavelengths of 700–1100. Total soluble solid (TSS) and TA were then calculated using a digital refractometer. Prior to calibration, the spectral data were translated using a number of mathematical methods and prediction models based on partial least square regression were produced. In an effort to identify physiological issues, mango fruits were kept in a chamber at a temperature of 5 °C–1 °C to mimic the symptoms of a chilling injury [77].

1.7 Conclusion

The quality of a fruit lies in its nutritional constituents as well as its appearance, which both play a major role in determining its market value and the costumer's preference. Although fruit quality can be evaluated using several techniques, spectroscopic methods have many advantages when compared to others. For example, spectroscopic methods are simple and do not damage the fruit during the process, they can also quantify more than one attribute at a time. However, even though they are quite useful techniques, they do not provide spatial information about the fruit. Many new technologies are being developed that use the basics of spectroscopy to build novel systems that are robust and able to evaluate quality more accurately.

References

[1] Sinha N K, Sidhu J, Barta J, Wu J and Cano M P (ed) 2012 *Handbook of Fruits and Fruit Processing* (New York: Wiley)

[2] Quebedeaux B and Bliss F A 1988 Horticulture and human health. Contributions of fruits and vegetables *Proc. 1st Intl. Symp. Hort. and Human Health.* (Englewood, NJ: Prentice Hall)

[3] Prior R L and Cao G 2000 Antioxidant phytochemicals in fruits and vegetables; diet and health implications *HortScience* **35** 588–92

[4] Lattimer J M and Haub M D 2010 Effects of dietary fibre and its components on metabolic health *Nutrients* **2** 1266–89

[5] New S 2001 Fruit and vegetable consumption and skeletal health: is there a positive link? Nutrition Foundation *Nutr. Bull.* **26** 121–5

[6] Agte V and Gite S 2014 Diabetic cataract and role of antiglycating phytochemicals In *Handbook of Nutrition, Diet and the Eye* ed V R Preedy 1st (Burlington: Elsevier Inc.) pp 131–40

[7] Southon S 2000 Increased fruit and vegetable consumption within the EU: potential health benefits *Food Res. Int.* **33** 211–7

[8] Slavin J L and Lloyd B 2012 Health benefits of fruits and vegetables *Adv. Nutr.* **3** 506–16

[9] Payne M E, Steck S E, George R R and Steffens D C 2012 Fruit, vegetable, and antioxidant in-takes are lower in older adults with depression *J. Acad. Nutr. Diet.* **112** 2022–7

[10] Castejon M G and Casado A R 2011 Dietary phytochemicals and their potential effects on obesity: a review *Pharm. Res.* **64** 438–55

[11] Alinia S, Hels O and Tetens I 2009 The potential association between fruit intake and body weight—a review. International Association for the Study of Obesity *Obes. Rev.* **10** 639–47

[12] Pessu P O, Agoda S, Isong I U and Ikotun I 2011 The concepts and problems of postharvest food losses in perishable crops *Afr. J. Food Sci.* **5** 603–13

[13] PROFAV 2011 Promotion of fruit and vegetables for health *African Regional Workshop Arusha* Tanzania: 26–30 September

[14] *Global Environment Monitoring System (GEMS/Food)* 2012 (World Health Organisation) 2015

[15] Wurr D C E, Fellows J R and Phelps K 1996 Investigating trends in vegetable crop response to increasing temperature associated with climate change *Sci. Hortic.* **66** 255–26

[16] Woolf A B and Ferguson I B 2000 Postharvest responses to high fruit temperatures in the field *Postharvest Biol. Technol.* **21** 7–20

[17] Mauzerall D L and Wang X 2001 Protecting agricultural crops from the effects of tropospheric ozone exposure: reconciling science and standard setting in the United States, Europe, and Asia *Annu. Rev. Energy Env.* **26** 237–68

[18] Sekse L 1995 Fruit cracking in sweet cherries (*Prunu savium* L.). Some physiological aspects —a mini review *Sci. Hortic.* **63** 135–41

[19] Murray X J, Holcroft D M, Cook N C and Wand S J E 2005 Postharvest quality of 'Laetitia' and 'Songold' (*Prunus salicina* Lindell) plums as affected by preharvest shading treatments *Postharvest Biol. Technol.* **37** 81–92

[20] Devleesschauwer B, Marvasi M, Giurcanu M C, Hochmuth G J, Speybroeck N, Havelaar A H and Teplitski M 2017 High relative humidity pre-harvest reduces post-harvest proliferation of Salmonella in tomatoes *Food Microbiol.* **66** 55–63

[21] Ramjan M D and Ansari M T 2018 Factors affecting of fruits, vegetables and its quality *J. Med. Plants* **6** 16–8

[22] Alford D V 2007 *Pests of Fruit Crops: A Color Handbook* (Elsevier)
[23] García J L, Ruiz-Altisent M and Barreiro P 1995 Factors influencing mechanical properties and bruise susceptibility of apples and pears *J. Agric. Eng. Res.* **61** 11–7
[24] Ozdemir F and Topuz A 2004 Changes in dry matter, oil content and fatty acids composition of avocado during harvesting time and post-harvesting ripening period *Food Chem.* **86** 79–83
[25] Medlicott A P, Reynolds S B, New S W and Thompson A K 1988 Harvest maturity effects on mango fruit ripening *Tropical Agric.* (*Trinidad and Tobago*)
[26] Hancock J, Callow P, Serçe S, Hanson E and Beaudry R 2008 Effect of cultivar, controlled atmosphere storage, and fruit ripeness on the long-term storage of highbush blueberries *HortTechnology* **18** 199–205
[27] Elshiekh F A and Abu-Goukh A A 2008 Effect of harvesting method on quality and storability of grapefruits *U. KJ Agric. Sci.* **16** 1–14
[28] Alavi N and Mazloumzadeh S M 2012 Effect of harvesting and drying methods of seedless barberry on some fruit quality *J. Saudi Soc. Agri. Sci.* **11** 51–5
[29] Cai Y, Takeda F, Foote B and DeVetter L W 2021 Effects of machine-harvest interval on fruit quality of fresh market northern highbush blueberry *Horticulturae* **7** 245
[30] Aliasgarian S, Ghassemzadeh H R, Moghaddam M and Ghaffari H 2013 Mechanical damage of strawberry during harvest and postharvest operations *World Appl. Sci. J.* **22** 969–74
[31] Hussein Z, Fawole O A and Opara U L 2020 Harvest and postharvest factors affecting bruise damage of fresh fruits *Hortic. Plant J.* **6** 1–13
[32] Bof M J, Laurent F E, Massolo F, Locaso D E, Versino F and García M A 2021 Bio-packaging material impact on blueberries quality attributes under transport and marketing conditions *Polymers* **13** 481
[33] Stefaniak J, Sawicka M, Krupa T, Latocha P and Łata B 2017 Effect of kiwiberry pre-storage treatments on the fruit quality during cold storage *Zemdirbyste-Agri.* **104**
[34] Murthy D S, Gajanana T M, Sudha M and Dakshinamoorthy V 2009 Marketing and post-harvest losses in fruits: its implications on availability and economy *Indian J. Agri. Econ.* **64**
[35] Zhang H, Huang J, Li T, Wu X, Svanberg S and Svanberg K 2014 Studies of tropical fruit ripening using three different spectroscopic techniques *J. Biomed. Opt.* **19** 067001
[36] Oliveira-Folador G, Bicudo M D O, de Andrade E F, Renard C M G C, Bureau S and de Castilhos F 2018 Quality traits prediction of the passion fruit pulp using NIR and MIR spectroscopy *LWT* **95** 172–8
[37] Wang H, Peng J, Xie C, Bao Y and He Y 2015 Fruit quality evaluation using spectroscopy technology: a review *Sensors* **15** 11889–927
[38] Gogineni R and Chaturvedi A 2020 Hyperspectral image classification *Processing and Analysis of Hyperspectral Data*
[39] dos Santos Neto J P, de Assis M W D, Casagrande I P, Cunha Júnior L C and de Almeida Teixeira G H 2017 Determination of 'Palmer' mango maturity indices using portable near infrared (VIS-NIR) spectrometer *Postharvest Biol. Technol.* **130** 75–80
[40] McClure W F 1994 Near-infrared spectroscopy. The giant is running strong *Anal. Chem.* **66** 43A–53A
[41] Pissard A, Fernández Pierna J A, Baeten V, Sinnaeve G, Lognay G, Mouteau A, Dupont P, Rondia A and Lateur M 2012 Non-destructive measurement of vitamin C, total polyphenol and sugar content in apples using near-infrared spectroscopy *J. Sci. Food Agric.* **93** 238–44

[42] Martinsen P and Schaare P 1998 Measuring soluble solids distribution in kiwifruit using near-infrared imaging spectroscopy *Postharvest Biol. Technol.* **14** 271–81

[43] Mendoza F, Lu R, Ariana D, Cen H and Bailey B 2011 Integrated spectral and image analysis of hyperspectral scattering data for prediction of apple fruit firmness and soluble solids content *Postharvest Biol. Technol.*

[44] Muhua L, Peng F and Renfa C 2007 Non-destructive estimation peach SSC and firmness by mutispectral reflectance imaging *N.Z. J. Agric. Res.* **50** 601–8

[45] Williams P C and Sobering D 1993 Comparison of commercial near infrared transmittance and reflectance instruments for analysis of whole grains and seeds *J. Near Infrared Spectrosc.* **1** 25–32

[46] McCraig T N, McLeod J G, Clarke J M and DePauw R M 1993 Measurement of durum pigment with a near-infrared instrument operating in the visible range *Cereal Chem.* **69** 671–2

[47] Bureau S, Ruiz D, Reich M, Gouble B, Bertrand D, Audergon J M and Renard C M 2009 Rapid and non-destructive analysis of apricot fruit quality using FT-near-infrared spectroscopy *Food Chem.* **113** 1323–8

[48] Costa C, Antonucci F, Pallottino F, Aguzzi J, Sun D W and Menesatti P 2011 Shape analysis of agricultural products: a review of recent research advances and potential application to computer vision *Food Bioprocess Technol.* **4** 673–92

[49] Cubero S, Aleixos N, Moltó E, Gómez-Sanchis J and Blasco J 2010 Advances in machine vision applications for automatic inspection and quality evaluation of fruits and vegetables *Food Bioprocess Technol.* **4** 487–504

[50] Feng L, Wu B, Zhu S, He Y and Zhang C 2021 Application of visible/infrared spectroscopy and hyperspectral imaging with machine learning techniques for identifying food varieties and geographical origins *Front. Nutr.* **8**

[51] Pudełko A, Chodak M, Roemer J and Uhl T 2020 Application of FT-NIR spectroscopy and NIR hyperspectral imaging to predict nitrogen and organic carbon contents in mine soils *Measurement* **164** 108117

[52] Shrestha S, Knapič M, Žibrat U, Deleuran L C and Gislum R 2016 Single seed near-infrared hyperspectral imaging in determining tomato (*Solanum lycopersicum L.*) seed quality in association with multivariate data analysis *Sensors Actuators* B **237** 1027–34

[53] Liu Z and Jing W 2012 Hyperspectral endmember detection method based on bayesian decision theory *Adv. Intell. Soft Comput.* 727–32

[54] Wu D and Sun D W 2013 Advanced applications of hyperspectral imaging technology for food quality and safety analysis and assessment: a review—Part I: fundamentals *Innov. Food Sci. Emerg. Technol.* **19** 1–14

[55] Chen Y, Deng J, Wang Y, Liu B, Ding J, Mao X, Zhang J, Hu H and Li J 2013 Study on discrimination of white tea and albino tea based on near-infrared spectroscopy and chemometrics *J. Sci. Food Agric.* **94** 1026–33

[56] Xing J, Saeys W and de Baerdemaeker J 2007 Combination of chemometric tools and image processing for bruise detection on apples *Comput. Electron. Agric.* **56** 1–13

[57] Tahmasbian I, Morgan N K, Hosseini Bai S, Dunlop M W and Moss A F 2021 Comparison of hyperspectral imaging and near-infrared spectroscopy to determine nitrogen and carbon concentrations in wheat *Remote Sens.* **13** 1128

[58] Li X, Li R, Wang M, Liu Y, Zhang B and Zhou J 2018 Hyperspectral imaging and their applications in the nondestructive quality assessment of fruits and vegetables *Hyperspectral Imaging in Agriculture, Food and Environment*

[59] Engel J, Gerretzen J, Szymańska E, Jansen J J, Downey G, Blanchet L and Buydens L M 2013 Breaking with trends in pre-processing? *TrAC, Trends Anal. Chem.* **50** 96–106

[60] Nicolaï B M, Beullens K, Bobelyn E, Peirs A, Saeys W, Theron K I and Lammertyn J 2007 Nondestructive measurement of fruit and vegetable quality by means of NIR spectroscopy: a review *Postharvest Biol. Technol.* **46** 99–118

[61] Barnes R J, Dhanoa M S and Lister S J 1989 Standard normal variate transformation and de-trending of near-infrared diffuse reflectance spectra *Appl. Spectrosc.* **43** 772–7

[62] Maleki M, Mouazen A, Ramon H and de Baerdemaeker J 2007 Multiplicative scatter correction during on-line measurement with near infrared spectroscopy *Biosystems Eng.* **96** 427–33

[63] Barnes R, Dhanoa M and Lister S 1993 Correction to the Description of Standard Normal Variate (SNV) and De-Trend (DT) transformations in *practical spectroscopy with Applications in Food and Beverage Analysis—2nd Edition J. Near Infrared Spectrosc.* **1** 185–6

[64] Araújo M C U, Saldanha T C B, Galvão R K H, Yoneyama T, Chame H C and Visani V 2001 The successive projections algorithm for variable selection in spectroscopic multi-component analysis *Chemometr. Intell. Lab. Syst.* **57** 65–73

[65] Wu D, Sun D W and He Y 2012 Application of long-wave near infrared hyperspectral imaging for measurement of color distribution in salmon fillet *Innov. Food Sci. Emerg. Technol.* **16** 361–72

[66] Li H, Liang Y, Xu Q and Cao D 2009 Key wavelengths screening using competitive adaptive reweighted sampling method for multivariate calibration *Anal. Chim. Acta* **648** 77–84

[67] Cai W, Li Y and Shao X 2008 A variable selection method based on uninformative variable elimination for multivariate calibration of near-infrared spectra *Chemometr. Intell. Lab. Syst.* **90** 188–94

[68] Centner V, Massart D L, de Noord O E, de Jong S, Vandeginste B M and Sterna C 1996 Elimination of uninformative variables for multivariate calibration *Anal. Chem.* **68** 3851–8

[69] Bureau S, Scibisz I, le Bourvellec C and Renard C M G C 2012 Effect of sample preparation on the measurement of sugars, organic acids, and polyphenols in apple fruit by mid-infrared spectroscopy *J. Agric. Food Chem.* **60** 3551–63

[70] Angra S K, Dimri A K and Kapur P 2009 Nondestructive brix evaluation of apples of different origin using near infrared (nir) filter based reflectance spectroscopy *Instrum Sci. Technol.* **37** 241–53

[71] Luo X, Takahashi T, Kyo K and Zhang S 2012 Wavelength selection in vis/NIR spectra for detection of bruises on apples by ROC analysis *J. Food Eng.* **109** 457–66

[72] Pissard A, Pierna J A F, Baeten V, Sinnaeve G, Lognay G, Mouteau A, Dupont P, Rondia A and Lateur M 2013 Non-destructive measurement of vitamin c, total polyphenol and sugar content in apples using near-infrared spectroscopy *J. Sci. Food Agric.* **93** 238–44

[73] Xia J, Li X, Li P, Wang W and Ding X 2007 Approach to nondestructive measurement of vitamin C content of orange with near-infrared spectroscopy treated by wavelet transform *Trans. CSAE* **23** 170–4

[74] Zude M, Herold B, Roger J M, Bellon-Maurel V and Landahl S 2006 Non-destructive tests on th prediction of apple fruit flesh firmness and soluble solids content on tree and in shelf life *J. Food Eng.* **77** 254–60

[75] M. Hussin M A, A Wahid M, M. Ahmad Hambali N A, Shahimin M M, Hasanuddin N and M. Azidin M 2017 Absorbance and transmittance capability of mangoes, grape and orange in NIR region *MATEC Web Conf.* **97** 01058

[76] Pronprasit R and Natwichai J 2013 Prediction of mango fruit quality from NIR spectroscopy using an ensemble classification *Int. J. Comput. Appl.* **83** 25–30
[77] Theanjumpol P, Self G, Rittiron R, Pankasemsuk T and Sardsud V 2014 Quality control of mango fruit during postharvest by near infrared spectroscopy *Chiang Mai Univ. J. Nat. Sci.* **13**

IOP Publishing

Advances in Spectroscopic Analysis of Food and Drink

Ashutosh Kumar Shukla

Chapter 2

Irradiation-based spectroscopic methods for quality analysis of fruits

Grzegorz Piotr Guzik and Wacław Stachowicz

Fruits that are treated with ionizing radiation can maintain their quality and freshness for longer. Irradiated fruits can be indetified with detection methods based on EPR/ESR spectroscopy. This test study aims to verify the suitability of detection methods available in the market to identify irradiated fruits. It will also discuss some novel aspects of EPR detection of irradiated fruits containing cellulose.

2.1 Introduction

A considerable volume of fruits are now imported. All fresh and dried food selected for export undergoes an obligatorily disinfection procedure to meet general sanitary regulations instituted by internationally authorized bodies to prevent the transfer of parasitic insects and microorganisms that are the source of dangerous foodborne diseases spread in the human population. One of the disinfection procedures suitable and adapted to fresh and dried fruit is ionizing radiation, such as with gamma rays emitted from commercial ^{60}Co irradiation sources or with the beams of high energy electrons generated from powerful electric accelerators. Radiation treatment is considered by the WHO/FAO board of experts to be an effective method of pasteurization and sterilization of food and food commodities. The advantage of irradiation is that it saves the fresh appearance and texture of food after radiation treatment, and effectively kills parasitic insects and pathogenic microorganisms that are found on the surface and inside of foods.

It has been confirmed experimentally and proven through analytical practice that freshness and high quality are essential features of fruits taken for radiation treatment, influencing the satisfactory result of food disinfection and/or microbial decontamination by means of ionizing radiation. Some time ago efforts were undertaken to improve the quality of fruits kept stored for a prolonged time by

doi:10.1088/978-0-7503-5573-5ch2 © IOP Publishing Ltd 2023

applying sophisticated combined irradiation techniques, such as the application of controlled portions of gamma rays or accelerated electrons and changing the air atmosphere over the irradiated fruits with nitrogen. However, using noble gases or gaseous compositions, varying the humidity, and so on did not deliver any positive results, either in the look or then texture of fruits after irradiation. It has been observed in contrast that neither regular nor combined radiation treatment on second class fruits stimulates undesired rotting of the latter during the storage. No wonder then that freshness and high quality of food taken for preservation by means of ionizing radiation is strongly endorsed in all FAO/WHO normative documents concerning food irradiation [1–3]. The term Good Irradiation Practice (GIP) related to radiation-treated food when correctly processed is also a sign of the high quality and freshness of a product. Consequently, any single fruit or batch of fruits labeled irradiated may be deemed to be a high-quality product of extended suitability for consumption.

2.2 Irradiated fruits

Global fruit turnover is dominated by fresh and dried fruits, the latter produced traditionally in Africa and Asia by keeping fresh fruits in the open air for a prolonged period, resulting in gradual dehydration. This primitive drying procedure protects fruits from disintegration during storage or transportation but does not limit the propagation of harmful insects and microbial contaminants responsible for foodborne diseases. For that reason, traditionally dried fruits undergo obligatory sanitary processing, preferentially with the use of ionizing radiation. An FAO/WHO expert group [2, 3] established the range of doses of ionizing radiation suitable for food to range from 0.2 to 10.0 kGy, depending on the target of the treatment chosen. Low doses, for example, are used to avoid germination and unwanted development of fruits and vegetables, middle doses are used to avoid insect propagation, and higher doses are used for microbial decontamination and sterilization of food [3]. Disinfection of fresh fruits requires the application of doses from 0.5 to 2.0 kGy, while microbial decontamination ranges from 3 to 7 kGy. Radiation sterilization of food dishes for special care patients and cosmonauts requires irradiation with a dose higher than 30 kGy.

2.3 Spectroscopic methods for identification of irradiated fruits

Analytical control of fruits treated with ionizing radiation is not a simple task. Classic and modern, and more or less sophisticated methods of analysis for the detection of accidental contaminants in food, as well as methods for the evaluation of food quality, are useless in this respect. The problem is that the distribution of radiation in food products appears in negligible concentrations. It has been evaluated that the treatment of food with 1 kGy of gamma radiation results, on average, in the production of ca 6 new molecular products per each of 10^6 molecules unchanged. The development of food irradiation technology and the increasing number of food irradiation facilities implies that studies for the detection methods for irradiated food are required. The necessity of detection methods to identify irradiated food was accentuated by the recommendation formulated by the board of authorized experts representing governments of a hundred countries in the *Conference of Acceptance Control and Trade in Irradiated Food*, which was held in 1988 in Geneva [4]. Extensive studies undertaken in many countries have resulted in the submission of more than 20 proposals for analytical methods capable of detecting irradiated food to be considered by the internationally authorized board of the Committee for Normalization in Brussels (CEN). Selected methods were undertaken in inter-laboratory testing on an international scale. Finally, 10 methods obtained the positive opinion of CEN experts and were issued in the form of European Standards recommended for general use. Currently, European Standards for the detection of irradiated food mentioned are authorized in many counties. Six of the 10 authorized detection methods are spectroscopic. Two popular detection methods for irradiated food based on thermoluminescence [5] and photoluminescence [6] measurements of isolated mineral impurities [5, 6] were found to be inconvenient for the detection of irradiated fruits, which are virtually free of the minerals mentioned.

The crux of any analytical method suitable for the detection of irradiated fruits lies in the ability to identify very low concentrations (ca 10^{-11} mol) of radicals or any entity with unpaired electron produced by radiation. EPR/ESR spectroscopy is the only method capable of meeting this requirement. Consequently, four of the 10 CEN standardized methods for detection of irradiated foods issued are based on EPR spectroscopy. Two of these methods are suitable for identification of fruits treated with ionizing radiation, and will be discussed in detail in this chapter.

One of the standardized detection methods recommended for general use is addressed to fresh fruits containing cellulose [7], which is the component of fiber containing constituents of fruits, such as acini, pips and stones, or peels. Cellulose is defined chemically as a linear polysaccharide that is composed of thousands of glucose links organized in semi-crystalline structures. The exposure of fruits to the action of ionizing radiation is immediately followed by the precise penetration of high energy gamma rays emitted from ^{60}Co sources inside fruits. Their constituent containing cellulose is the target for the formation of radiation-induced radicals remaining incorporated with cellulose molecule but giving rise in EPR to isotropic triplet with 1:3:1 line proportion distanced 6 mT.

Figures 2.1 and 2.2 demonstrate working examples of EPR records obtained with pips taken from kiwi fruit and red paprika irradiated technologically with an unknown dose of ionizing radiation. The recognizable specificity of EPR signal proving radiation treatment of investigated fruit sample lies in its symmetry, highly intense central line, and first of all in the pair of distinguishable side lines distanced 6 mT representing an overlapped radiation-induced isotropic triplet.

The second of two standardized methods suitable for the detection of irradiated fruits is addressed to dried fruits containing crystalline sugar [8]. The finding of crystalline sugar in fresh fruit is less probable but is common and natural in dried fruits. The formation of crystalline sugar spots in fruits intentionally dried occurs as the consequence of the dehydration of fruit, which consists in the gradual thickening of saturated syrup accumulated inside the intra-fibroid domains of the fruit.

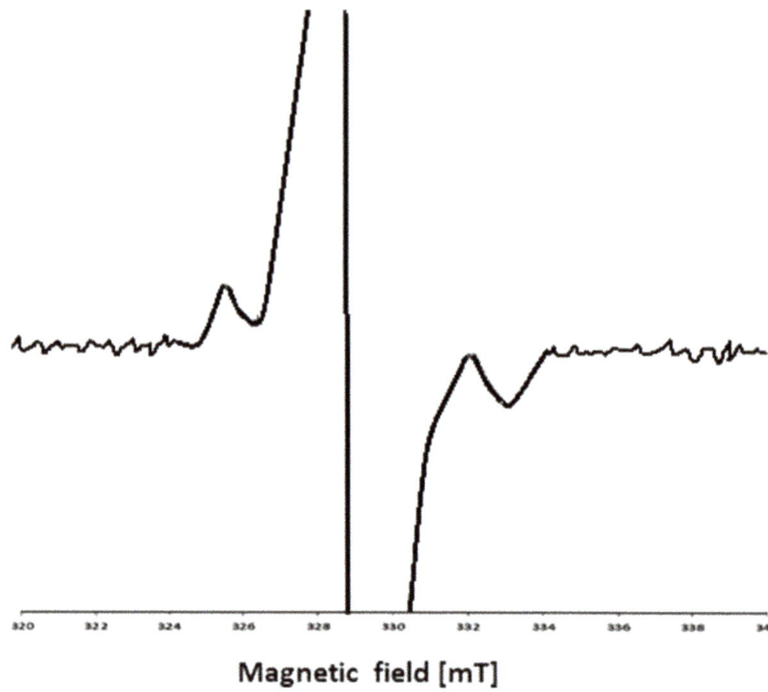

Figure 2.1. First derivative EPR signal of fresh *red paprika* pips irradiated with 4 kGy of ^{60}Co gamma radiation. Low and high magnetic field side EPR lines of radiation-induced cellulose radical are well distinguished.

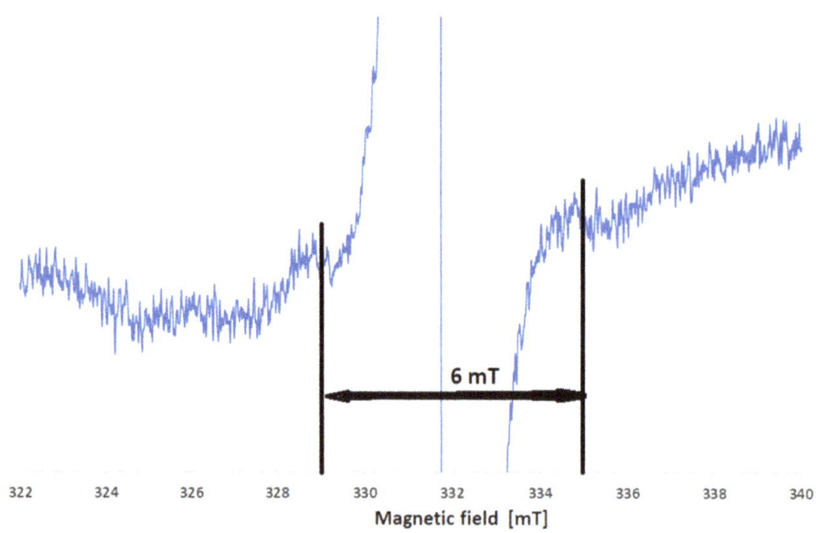

Figure 2.2. First derivative EPR signal of *kiwi pips* irradiated with 4 kGy of gamma radiation. Two low and high magnetic field side satellite lines of cellulose radical distanced 6 mT are well distinguished, despite the signal noise level.

During irradiation, the incident gamma rays penetrate the interior of the fruit and sugar spots suspended in saturated juice and interact preferentially with crystalline lattice imperfections of sugar spots. This results in the formation of sugar-originated radicals trapped durably inside the sugar spots. In EPR, this gives rise to a signal that is defined as multiline and is specifically broad, as shown in figure 2.3 recorded with an irradiated resin sample. Figure 2.4 presents a similarly-shaped sugar-originated EPR signal derived from the fresh flesh of a banana. This is perhaps the first observation of a radiation-induced sugar-originated EPR signal derived from fresh fruit. It is worth noting at this point that there some dried fruits, such as plum, apricot, and berries, are free of crystalline sugar, despite the depth of the dehydration process.

Multiline EPR signal recoded with dried fruits exposed to the action of ionizing radiation is defined sometimes by the envelope of the EPR signals derived from radiation-induced radicals originated from numerous sugars and their derivatives contained in the fruit. A sugar-derived multiline EPR signal is characterized with g_c (g central) equal to 2.005, which is the number established 'by weight' as the point between positive and negative areas of the record of the first derivative EPR signal registered. The second specific feature of EPR signal derived from radiation-induced sugar-originated radicals is its broadness of 8–9 mT. In contrast to EPR signals derived from radiation-induced cellulose radicals, the discussed sugar-originated EPR signals are free of native signals of organic origin. It is presumed that native signals of this kind are temperature sensitive and decay by drying.

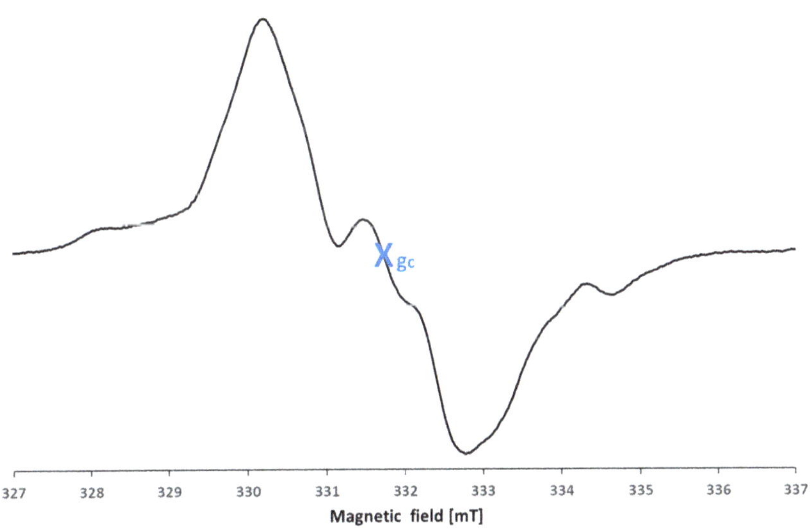

Figure 2.3. First derivative multiline EPR signal recorded with *sultan raisins* irradiated with 4 kGy of gamma radiation, representing a family of EPR signals assigned to irradiated fruits containing crystalline sugar, glucose, and fructose, preferentially.
$g_{central} = 2.005$; EPR signal width = 9.5 mT.

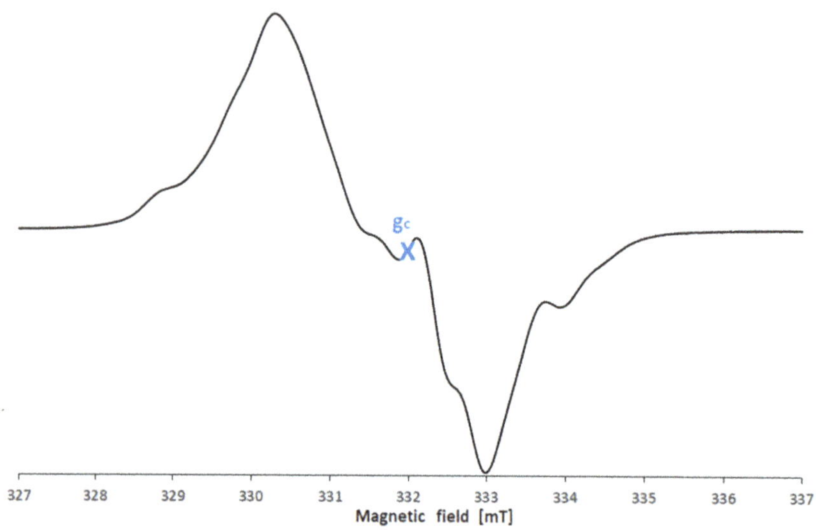

Figure 2.4. First derivative multiline EPR signal recorded with *banana pulp* taken from fresh banana fruit irradiated with 4 kGy of gamma radiation. It is, to our knowledge, the first example of finding crystalline sugar in fresh fruit.
$g_{central}$ = 2.005; EPR signal width = 7.8 mT

2.4 Study of the fruit constituents that are suitable for the detection of irradiated fruits

An independent test study has been carried out to verify the suitability of standardized detection methods for the identification of radiation treatment on fresh fruits containing cellulose [7] and those containing crystalline sugar [8] in practice. Fresh and dried fruits were purchased in an open-air market taken by chance from a market stand. The choice of fresh fruits included strawberries, kiwi fruit, pear, apple, watermelon, orange, tangerine, banana, tomatoes, and red paprika, while that of dried fruits sultana raisins, pineapple, fig, and vanilla flower. Sample preparation and EPR measurements were conducted based on European Standards EN 1787: 2001 [7] and EN 13708: 2003 [8], respectively. Once the fruits were irradiated with 4 kGy of ^{60}Co gamma rays and kept overnight in dark at room temperature, the components indicated cellulose and sugar reach were isolated from the fruits, preferentially from the pips and peels if accessible. The components excised from fresh fruits were cut into smaller pieces and placed in EPR measuring ampoules. Operational parameters for EPR measurements done were consistent with those recommended in the related European Standards [7, 8].

Unexpectedly, among the 10 samples of fresh fruits examined only five listed in table 2.1 were classified as positively irradiated meeting the requirements of European Standard 1787:2001 concerning the distance between two outer lines of radiation-induced EPR signal.

Table 2.1. Irradiated fruit components containing cellulose radicals giving rise in EPR to isotropic triplet of spectral characteristics consistent with the requirement of European Standard EN-1787:2001.

Number	Fruit component examined	Distance between outer lines of EPR triplet derived from radiation-induced cellulose radial (see also figure 2.1)	g_o[a] of radiation-induced EPR signal	g_o of native EPR signal recorded
1	Vanilla flower lobule dried	6.0±0.3	2.004	2.004
2	Red paprika pips fresh	6.0±0.3	2.003	2.004
3a	Kiwi peel fresh	6.0±0.4	2.003	2.004
3b	Kiwi pips fresh	6.0±0.1	2.005	2.004
4a	Pear peel fresh	6.0±0.1	2.004	2.004
4b	Pear pips fresh	6.0±0.4	2.005	2.004
5	Watermelon pips fresh	6.0±0.4	2.003	2.004

[a] g_o—spectroscopic splitting factor.

Table 2.2. Irradiated fruit constituents containing cellulose giving rise in EPR to isotropic triplet of spectral characteristic differing from that required in European Standard EN 1787:2001.

Number	Fruit constituent EPR examined	Distance between outer lines of radiation-induced EPR triplet	g_o of radiation-induced EPR signal	g_o[a] of native EPR singlet recorded
1a	Tangerine peel	7.4	2.005	1.999
1b	Tangerine pips	8.2	2.004	2.000
2a	Orange stone	7.0	2.005	2.001
2b	Orange peel	7.7	2.004	2.001
3a	Apple pips	8.3	2.003	2.000
3b	Apple peel	7.0	2.003	2.000
4a	Tomato pips	7.3	2.004	2.000
4b	Tomato peel	7.3	2.004	2.001
5	Watermelon peel	8.9	2.004	2.001

[a] g_o—spectroscopic splitting factor.

The remaining five samples of fresh fruits containing cellulose are compiled separately in table 2.2 as not qualified, as irradiated, or not irradiated in view of European Standard EN-1787:2001. None of the five samples had an EPR measurement that met the absolute requirement stated in the standard by recording an EPR signal at a distance 6 mT between two left- and right-hand side lines of a radiation-induced isotropic triplet, derived presumably from induced cellulose radical radiation. In such a position, the discussed samples could only be qualified as unclassified, despite the fact that they were characterized with stable radiation-induced EPR signals, resembling those obtained with samples meeting the standard requirement and classified as radiation treated.

The problem will be demonstrated and commented on by taking advantage of the experimental results obtained with irradiated tangerine pips, as shown in figure 2.5.

The EPR signal recorded with irradiated tangerine pips (red line) represents an isotropic triplet resembling that obtained with fruit samples meeting the requirement of EN 1787:2001 standard, as shown in table 2.1. The difference lies in the splitting between individual lines of radiation-induced triplet, enhanced now to 8.2 mT and being markedly higher than the 6.0 mT required in European Standard. A similar enhancement of EPR line splitting ΔH has also been observed with the next four samples of irradiated fruits, characterized this time with EPR line splitting higher than 6 mT but varying from 7.4 to 8.9 mT, as shown in table 2.2. The observed effect that is proven with five samples of irradiated fruits containing cellulose has neither been reported nor discussed in the literature to our knowledge. Looking at table 2.3, which compiles the data presently available on the contents of cellulose in popular fruits, it can be seen that cellulose concentration cannot be the only factor

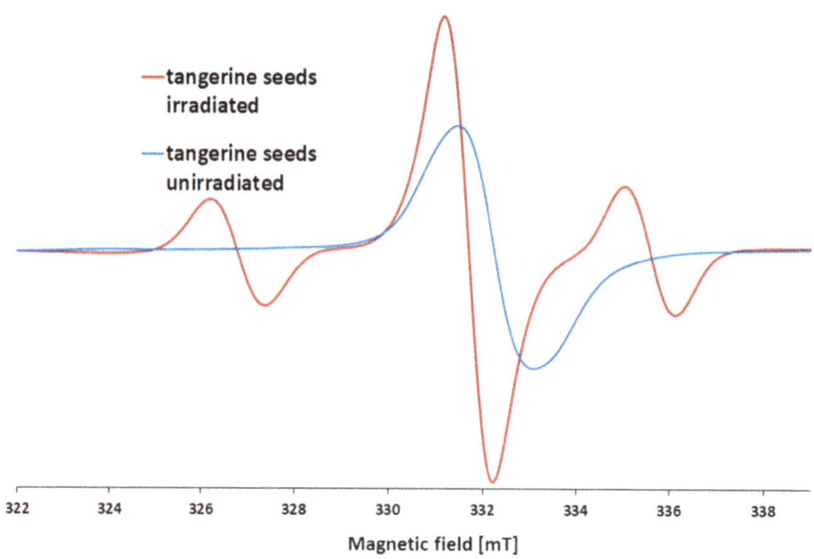

Figure 2.5. First derivative EPR signal of *tangerine pips* taken from fresh tangerine fruit irradiated with 4 kGy of ^{60}Co gamma radiation (red line) together with native signal of non-irradiated tangerine fruit (blue line). EPR signal of irradiated tangerine pips represent isotropic triplet partly overlapped with native one resulting in slight asymmetry of the latter. The distance between the left- and right-hand side lines of radiation-induced EPR triplet equals 8.2 mT and is markedly greater than the 6.0 mT required.

g_o estimated for radiation-induced triplet—2.004

g_o estimated for native singlet—2.000

Table 2.3. Cellulose content estimated in water insoluble[a] fiber fraction separated from popular fruits [10, 11].

Fruits	Insoluble fiber [g in 100g of fruits] (for example cellulose)
Apple	1.54
Avocado	3.51
Banana	1.21
Grapes	0.36
Orange	0.99
Tangerine	1.80
Peach	1.54
Plum	1.76
Prune	3.63
Raisins	2.17
Strawberries	1.80
Kiwi	2.10

[a] Fibers are composed of soluble and water-insoluble fractions. Water-insoluble fiber fraction contains cellulose, silica, and a lot of other ingredients [9, 12–14].

Table 2.4. Irradiated fruit components giving rise in EPR to a broad multiline signal assigned to sugar-originated radicals.

Number	Fruit constituent EPR examined	Total EPR signal width (mT)	Spectroscopic splitting factor g_c
1	Sultan raisins flesh dried	9.50	2.005
2	Fig flesh dried	9.20	2.004
3	Pineapple flesh dried	8.10	2.004
4	Fresh banana flesh	7.80	2.005

responsible for the enhancement of interline distance observed by spectral analysis of isotropic triplet radiation. It has to be recognized that cellulose is only one of many components of a fibroid matrix, such as a conglomerate of polysaccharides and various non-polysaccharide structured carbohydrates such as lignin. Rigid fiber itself is a coherent structure that is capable of trapping stable radical's radiation, which is not necessarily evoked by only cellulose originated [9]. However, it is still early to draw a binding conclusion concerning the newly observed phenomenon. The repeatability of EPR measurements or the radiation detection limit are basic questions that still need to be answered.

Table 2.4 compiles the results of EPR examination on dried fruits containing crystalline sugar that were exposed to the action of gamma radiation. The EPR signals obtained were defined as multiline, and were identically shaped and represented by the same basic spectral parameters as those considered specific for radiation-treated fruits in European Standard EN 13708:2002 [8].

Attention is proposed to be paid to position no. 4 related to banana flesh taken from fresh fruit and examined as such. The positive identification of radiation treatment on fresh banana fruit with the use of a standardized detection method addressed preferentially to dried fruit is a valuable achievement [8].

It should be kept in mind that the stability of the recorded EPR signal [15, 16] is an indispensable requirement for any detection method for irradiated fruits applicable in practice [17, 18]. Consequently, the EPR signals recorded with samples of fruits that were used to conduct the described test experiment have been preliminarily controlled. It has been proven that neither structural change nor intensity reduction of radiation-induced EPR signal recorded with investigated fruits were observed after storage at room temperature for three months from the day of treatment. It is pertinent to note that irradiated dried fruits belong to the assortment of foodstuffs in which a radiation-induced EPR signal survives unchanged for a longer period than that recorded with fresh, chilled, or refrigerated foodstuffs.

References

[1] Directive 1999 /2/EC of the European Parliament and of the Council on the approximation of the laws of the member states concerning foods and food ingredients treated with ionizing radiation.

[2] Directive 1999 /3/EC of the European Parliament and of the Council on the establishment of a Community list of foods and food ingredients treated with ionizing radiation.

[3] FAO/WHO General standard for irradiated foods *Codex Stan 106–1983, Rev.* 1–2003

[4] International Consultative group on Food Irradiation 1991 *Code of good irradiation practice for shelf-life extension of bananas, mangoes and papayas, ICGFI Document* No. 6 International Atomic Energy Agency

[5] European Committee for Standardization 2001 *Foodstuffs—thermoluminescence detection of irradiated food from which silicate minerals can be isolated* EN-1788 European Union

[6] European Committee for Standardization 2009 *Foodstuffs—luminescence detection of irradiated food from which silicate minerals can be isolated* EN-13751 European Union

[7] European Committee for Standardization 2001 *Foodstuffs—Detection of Food Containing Cellulose by ESR Spectroscopy* European Standard 1787 European Committee for Standardization

[8] European Committee for Standardization 2003 *Foodstuffs—Detection of Irradiated Food Containing Crystalline Sugar by ESR Spectroscopy* European Standard EN 13708 European Committee for Standardization

[9] Gerting H and Przysławski J 2006 Bromatology. An outline of food and nutrition science (In Polish only) *Warszawskie Wydawnictwo Lekarskie PZWL* p 37

[10] Dembiński Ł, Banaszkiewicz A and Radzikowski A 2010 High-residue diet—definition, benefits and norms in pediatrics *'Contemporary Pediatrics. Gastroenterology, Hepathology and Child Nutrition 12'* (In Polish only) pp 139–45 2

[11] Betty W L, Karen W A and Pamela R P 2002 Individual sugars, soluble, and insoluble dietary fibre contents of 70 high consumption foods *J. Food Compos. Anal.* **6** 715–23

[12] Bartnikowska E 1997 Dietary fibre in human nutrition. Part I *Food Ind.* **51** 43–8 (In Polish only)

[13] Hasik J 1997 The role of plant fibre in human nutrition. Warsaw (In Polish only) S G G W Wydawnictwo

[14] Świderski F (ed) 2003 Convenience food and functional food. Warsaw (In Polish only) *Wydawnictwo Naukowo-Techniczne* pp 278–86

[15] Raffi J and Agnel J-P 1989 Electron spin resonance identification of irradiated fruits *Radiat. Phys. Chem.* **34** 891–4

[16] Raffi J, Agnel J-P, Buscarlet L A and Martin C C 1988 Electron spin resonance identification of irradiated strawberry *J. Chem. Soc. Faraday Trans* **1** 3359–62

[17] Miller R B 2005 *Electronic Irradiation of Foods: An Introduction to the Technology* (New York: Springer) 295

[18] Bortolin E, Boniglia C, Gargiulo R and Onori S 2009 Herbal materials used in dietary supplements: comparison of luminescence methods for detection of irradiation *Radiat, Phys, Chem.* **78** 683–5

IOP Publishing

Advances in Spectroscopic Analysis of Food and Drink

Ashutosh Kumar Shukla

Chapter 3

Authenticity assessment of natural and artificial compounds in fruit juices by using FTIR

Prinya Wongsa

3.1 Introduction

The world's population was 7,922,312,800 people at the end of September 2022, and was expected to reach 8 billion by mid-November of 2022. The population continues to increase by roughly 140 people per minute [1]. Population growth is directly proportional to the demand and supply of raw materials and food products. The global food market revenue was around 9.43 trillion US in 2023 and is expected to increase by 4.79% annually till 2027. Moreover, the average volume per person in the food market is estimated to be 337.8 kg in 2022 [2]. In addition, customers are well-educated, concerned about quality seals and trust marks on food products, and they expect manufacturers and retailers to indicate the safety and high-quality of their products on the labels of packaged foods [3]. An increase in the market size of the food industry has resulted in the strict monitoring and implementation of quality analysis and quality management systems to maintain the quality, health benefits, and safety of food products. Quality control and assurance activities can be achieved by regular practice and involve all parties, including food production, processing, quality control, transportation, warehouse, and dealers [4].

3.2 Food quality and authenticity control

Food adulteration refers to the alteration of food quality that takes place deliberately. It includes the addition of ingredients to modify different properties of food products for economic advantage. Due to the increased popularity and high price of raw materials, e.g., some exotic super fruit juices, 100% juice can be adulterated with less costly alternatives, including water, sugar, or other juices that are less expensive to produce, which dilute the original juice without changing the taste [5].

doi:10.1088/978-0-7503-5573-5ch3

© IOP Publishing Ltd 2023

Food fraud and authenticity are important issues for the fruit juice market worldwide. Food adulteration is a primary global concern for public health, especially in developing countries, due to the lack of monitoring, and appropriate policy development and implementation. In addition, food adulteration is a significant challenge in food safety and regulation in many countries.

There is no a consensus definition of food fraud or economically motivated adulteration (EMA) of foods or food ingredients (a subset of food fraud), which can be defined as the intentional contamination of food for increased profit [6]. Food fraud is a collective term that encompasses the deliberate substitution, addition, tampering, or misrepresentation of food, food ingredients, or food packaging, or making false or misleading statements made about a product for economic gain [6, 7]. The US Food and Drug Administration (FDA) adopted a 'working definition' for an April 2009 public meeting to raise awareness and solicit public input regarding EMA of FDA-regulated products. The FDA defined 'economically motivated adulteration' as the fraudulent and intentional substitution or addition of a substance to a product for the purpose of increasing the apparent value of the product or reducing the cost of its production, i.e., for economic gain. EMA includes dilution of products with increased quantities of an already-present substance, e.g., increasing inactive ingredients of a drug with a resulting reduction in strength of the finished product or watering down of juice, to the extent that such dilution poses a known or possible health risk to consumers, as well as the addition or substitution of substances to mask dilution [8]. In addition, the European Union (EU) report on food fraud states the EU laws do not provide for a 'generally acknowledged definition of food fraud', despite an extensive legislative framework focused on food safety. The only general guideline is found in EU regulations requiring that food labeling, advertising, presentation, and packaging shall not mislead consumers [9].

3.3 Food fraud and authenticity

Food products can be frauded in the following ways:

First, replacement by complete or partial replacement of a food ingredient or valuable authentic composition with a less expensive substitute, which is usually achieved through the addition, dilution, or extension of an authentic ingredient with an adulterant or mixture of adulterants, e.g., the addition of water and citric acid to lemon juice to fraudulently increase the titratable acidity of the final juice product [10]. Moreover, this adulteration generates false claims and non-declarations including mislabeling and misrepresentation of packaging label messages for economic gain. False declaration of geographic, species, botanical, or varietal origin is also observed. An example is the substitution of Greek or Turkish olive oil labeled as Italian olive oil [11]. Fraudulent practices also cause a false declaration of the production process, e.g., fraudulent labeling of a synthetically derived flavor chemical as being naturally derived or food falsely labeled or certified as organically produces [12].

Second, the addition of small amounts of a non-authentic substance to mask an inferior quality ingredient, e.g., the addition of a color additive (such as Sudan Red dyes) to enhance the color of poor-quality palm oil [13].

Finally, the removal of an authentic and valuable constituent without the consumer noticing. One example is poor-quality honey that is filtered to remove pollen or other residue from the beehive to make it difficult to determine the honey's botanical and geographic origin [14].

In terms of authenticity, food fraud, or EMA sometimes give rise to authenticity issues, including the misrepresentation of the brand, origin, manufacturing ingredients, and their composition. Authenticity is defined in Webster's Dictionary as 'of undisputed origin, or to be genuine' [5]. In addition, many studies have also provided some of the familiar terms including Protected Designation of Origin (PDO), Protected Geographical Indication (PGI), Certificate of Specific Character (CSC), and Traditional Specialties Guaranteed (TSG) [5].

Food fraud can place consumers at immediate risk, they may suffer long-term harm, and it can create direct or indirect food safety risks. Preventive actions including assessment of food fraud risks, a set of preventive measures to manage risk, regular verification of measures, and training operators in the factory, which should be operated and monitored by food producers and manufacturers. Moreover, the consumers should carefully select the product, read the label, ingredients, and manufacture and expiration date before purchasing and consuming food products.

3.4 Detection of adulteration and authenticity

Food products, e.g., fruit-based products such as juices, jams, and purees, might be adulterated by blending these products with cheaper fruits. In addition to admixtures of adulterants, the labeled fruit contents may not be met. Consequently, adulteration is difficult to detect and can lead to poor product quality. For consumer protection and to avoid unfair trade, it is essential to assure both authenticity and compliance with the product specification. However, the heterogeneous and complex nature of food products are major challenges for analytical methods. Analytical techniques for food authenticity include both traditional and modern sophisticated instrumental techniques, such as proximate analysis, DNA-based approaches, chronographic, spectroscopic, and radiometric techniques. Consequently, suitable and applicable methods for authenticity control and determination of adulterants in foods products have been studied and reviewed [4]. In addition, the Association of Official Analytical Chemists (AOAC) has well documented the analytical procedures to determine the quality of fruit juices and soft drink [15]. The analysis of parameters of fruit juices is typically based on titrimetric or volumetric analyzes, including estimation of vitamin C and titratable acidity, which involve the extensive use of solvents and reagents [15]. In addition, chemical analysis of food samples to determine their composition and characteristics must be investigated to achieve products' quality standards, ensure compliance to food authority recommendations, and provide accurate nutritional labeling for processed food products. Li *et al* [16] reported a comparison of profiles of sugars and organic acids among different fruit juices (including apple, pear, peach, grape, sweet cherry, strawberry, and blueberry, with various varieties) to authenticate using chemometrics. In addition, DNA-based technologies, including polymerase chain reaction (PCR), analysis, microarrays, and next generation sequencing, are useful tools for a number of food authentication approaches. The PCR method was utilized for species determination in foods containing DNA in detectable amounts, particularly meat and fish [16]. The application of DNA and proteins to authenticate and quality control fruits and derived products has also been studied. For example, an accurate and robust method to identify both known and unknown fruit species in small berry fruit products using DNA barcoding technology based on Sanger sequencing has been adopted [17]. In addition, spectroscopic techniques, including mass spectrometry and nuclear magnetic resonance (NMR) spectroscopy, have been carried out to detect the adulteration of fruit juices. NMR targeted and untargeted methodologies are applicable tools for food authentication and the detection of counterfeit products. Targeted approaches allow the identification of specific markers of adulteration for food products. In the untargeted approach, the chemical profile of food materials is used to create a unique fingerprint as a reference for suspect samples [18]. Moreover, Isotope ratio mass spectrometry (IRMS) has been used to characterize the sugar-specific carbon isotope ration in coconut waters for authentication purposes [19].

Vibrational spectroscopy techniques, including near-infrared spectroscopy (NIR) and Fourier-transform mid-infrared (FT-MIR) spectroscopy, are simple, rapid, and non-destructive techniques. Applications of FT-MIR techniques include assessing the maturity or quality attributes of products, screening progeny from breeding programs, determining of plant metabolites and extent of adulteration, and authenticating products for type or geographic locations [20]. The detection of quality attributes of various liquid foods, including alcoholic beverages (red wines, rice wines, and beers), nonalcoholic beverages (juices, fruit vinegars, coffee beverages, and cola beverages), dairy products (milk and yogurt), and oils (vegetable, camellia, peanut, and virgin olive oils and frying oil) has been studied using NIR techniques [21]. Some green analytical methods based on vibrational spectroscopy, including IR spectroscopy, have recently become widely developed, validated, and applied for food and beverage quality control, and to provide reliable food fingerprinting [22, 23].

3.5 Vibrational spectroscopy

Infrared is based on vibrational spectroscopy, which is used to analyze and interpret the relationship between the interaction of light with matter (solids, liquids, or gases). The wavelength or frequency of the infrared spectrum is subdivided into three regions: near-infrared (NIR) (14 000–4000 cm^{-1} or 714–2500 nm), mid-infrared (MIR) (4000–400 cm^{-1} or 2500–25 000 nm), and far-infrared (400–10 cm^{-1} or 25 000–1×10^6 nm.) (figure 3.1). In addition, infrared spectroscopy measures the wavelength or the frequency that the sample absorbs, as well as the intensity of the absorptions of infrared radiation by a sample [4].

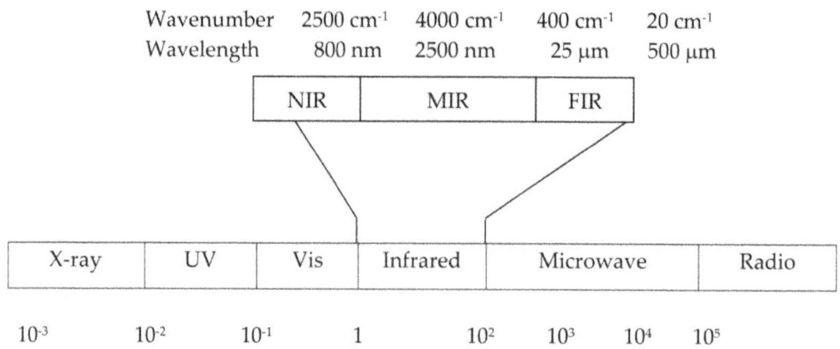

Figure 3.1. Infrared radiation regions.

3.6 Sample preparation for FTIR analysis

The preparation of solid foods starts by grinding the material to a fine powder and mulling in liquid paraffin (Nujol) or dispersing in an alkali halide disks made from potassium bromide (KBr). The theory behind the preparation of a powder sample with KBr is that a simple inorganic salt does not generate any vibrations in the MIR region [24]. Generally, approximately 1–3 mg of the ground sample is thoroughly mixed with about 350 mg of ground KBr [25]. The mixture is transferred to a die that is then placed in a suitable press and pressed at around 12,000 psi. Consequently, a fused disk can be introduced into the light beam of a spectrophotometer. The application of KBr disks is a practical method in some studies, e.g., the detection of wheat four adulteration with barley four [26].

The FTIR spectrum quality is affected by the quality of the disk. Moreover, reproducibility by using KBr disks resulting from the matrix ratios and the homogeneity need to be exactly the same for each sample. A disk that is too thin can be caused by too little powder, while using too much pressure for too long can cause the disk to break on removal from the die. Therefore, the disks and mulls must be prepared under precisely same circumstances to avoid any changes in path length, and thus response [25]. Furthermore, due to the mixing of the sample and KBr, this dried sample mixing technique is destructive method [27]. The drawbacks related to disk sample preparation and the spectral reproducibility have recently been addressed by applying the attenuated total reflectance (ATR) technique [24]. The use of ATR-FTIR spectroscopy allows for the collection of the spectra from solids, liquids, semi-solids, and thin films. In ATR-FTIR techniques, the sample is pressed against high refractive index crystals, usually made from germanium, diamond, zinc selenide (ZnSe), or thallium iodide/thallium bromide (KRS-5) [25]. Meanwhile, in ATR-FTIR spectroscopy the reflection technique is based on the IR light being reflected internally off the back surface of an internal reflection element with a high index refraction, which is in contact with the sample. The IR beam travels inside the crystal and a standing wave of radiation is generated, which is called the evanescent wave [25]. A sample in contact with the crystal can interact with the evanescent wave, absorb IR, absorb energy, and have its IR spectrum detected. The evanescent wave is attenuated by the sample's absorbance. Therefore, this technique is so-called ATR [25]. In addition, the application of ATR-FTIR spectroscopy is a general technique for studying food authenticity or adulteration [27].

3.7 Interpretation of infrared adsorption frequencies

IR spectroscopy is one of the molecular spectroscopies in which the absorption of energy leads to several excited molecular vibrational and rotational states to provide infrared spectra. In addition, an infrared spectrum represents a fingerprint spectral region of a sample, with absorption peaks that correspond to the frequencies of vibrations between the bonds of the atoms. Therefore, this fingerprint can be used to confirm the sample's identity [4]. Moreover, bands or peaks come from the absorption of IR beam by functional groups in the samples. Each band is characterized by wavenumbers to which functional groups absorb, peak intensity (strong, medium, weak), peak shapes (broad and sharp), and type of vibrations (symmetric and asymmetric stretching vibration, bending) [28].

Each functional group between atoms of a molecule presents a unique vibrational frequency, which can then be employed to determine which functional groups are present in a sample. When the effects of all the different functional groups are taken together, the result is a unique molecular 'fingerprint' that can be used to confirm the identity of a sample [29]. Detailed information about the infrared absorptions observed for various bonded atoms and groups can be found in the literature. A summary of typical infrared adsorption frequencies of a few typical classes of compounds is given in table 3.1.

3.7.1 Spectra of sugars

Glucose, fructose, and sucrose are the dominant sugars to be found in most fruit juices. The spectra of pure reference glucose and sucrose are used to demonstrate the interpretation spectral, as shown in figure 3.2.

It can be seen in figure 3.2 that the spectra of sucrose (red line) and glucose (blue line) display a broad stretching intense characteristic peak near 3580–3160 cm^{-1} due to hydroxyl stretching vibration of the polysaccharide fractions, and a weak C–H stretching vibration band in the region of 2880 cm^{-1}. The band towards 1620 cm^{-1} was attributed to stretching vibration of C=O in the protonated carboxylic acid [33]. Moreover, the band at 1620 cm^{-1} was due to the bound water. Interestingly, the fingerprint regions for sucrose and glucose regain were found between 1480 and 800 cm^{-1} [34].

3.7.2 Spectra of orange juices

Four orange juices were prepared from different raw materials using laboratory scale processing methods. Their individual spectra, vibrational peaks, and tentative band assignments are presented in figure 3.3. Glucose, fructose, and sucrose are the dominant sugars to be found in most fruit juices [33–35].

The ATR FT–IR spectra of freshly squashed mandarin juice (red) and freshly squashed orange juice (black) demonstrate that the two juices can be discriminated based on spectral features, the most prominent of which is the highlighted band at

Table 3.1. Typical infrared adsorption frequencies.

	Stretching vibrations		Bending vibrations	
Functional class	Range (cm^{-1})	Assignment	Range (cm^{-1})	Assignment
Saturated aliphatic (alkane/alkyl) group frequencies				
Methyl ($-CH_3$)	2970–2950	Methyl C–H asym. stretch	1470–1430	Methyl C–H asym bend
	2880–2860	Methyl C–H sym. stretch	1380–1370	Methyl C–H sym. bend
Methylene ($-CH_2$)	2935–2915	Methylene C–H asym. stretch	1485–1445	Methylene C–H bend
	2865–2845	Methylene C–H sym. stretch		
Methyne (–CH–)	2900–2880	Methyne C–H stretch	1350–1330	Methyne C–H bend
	1300–700	Skeletal C–C vibrations		
Olefinic (alkene) group frequencies				
C=C	1680–1620	Alkenyl C=C stretch		
	1625	Aryl-substituted C=C		
	1600	Conjugated C=C		
C–H	3095–3075 + 3040–3010	Terminal (vinyl) C–H stretch	970–960	*trans*-C–H out-of-plane bend
			700 (broad)	*cis*-C–H out-of-plane bend
Alcohol and phenols	3580–3650	O–H (free), sharp	1330–1340	O–H bending (in-plane)
	3200–3550	O–H (H-bonded), broad	650–670	O–H bending (out-of-plane)
	970–1250	C–O		
Aromatic ring (aryl) group frequencies/phenols				
C=C–C	1615–1580	Aromatic ring stretch	1225–950 (several)	Aromatic C–H in-plane bend
C=C–C	1510–1450	Aromatic ring stretch	900–670 (several)	Aromatic C–H out-of-plane bend
			2000–1660 (several)	Aromatic combination bands

(*Continued*)

Table 3.1. *(Continued)*

Functional class	Stretching vibrations: Range (cm^{-1})	Stretching vibrations: Assignment	Bending vibrations: Range (cm^{-1})	Bending vibrations: Assignment
Hydroxy compound group frequencies				
O–H	3570–3200 (broad)	Hydroxy group, H-bonded; O–H stretch		
	3400–3200	Normal 'polymeric' O–H stretch		
	3550–3450	Dimeric O–H stretch		
	3570–3540	Internally bonded O–H stretch		
	3645–3600 (narrow)	Non-bonded hydroxy group, O–H stretch		
	3640–3530[c]	Phenols, O–H stretch		
O–H			1410–1310	Phenols, O–H bend
			1330–1340	O–H bending (in-plane)
			650–670	O–H bending (out-of-plane)
C–O	~1200[d]	Phenol, C–O stretch		
Amines	3400–3500	N–H (1°-amines), 2 bands	1550–1650	NH_2 scissoring (1° amines)
	3300–3400	N–H (2°-amines)	660–900	NH_2 and N–H wagging (shift on H-bonding)
	1000–1250	C–N		

Source: [29–31]

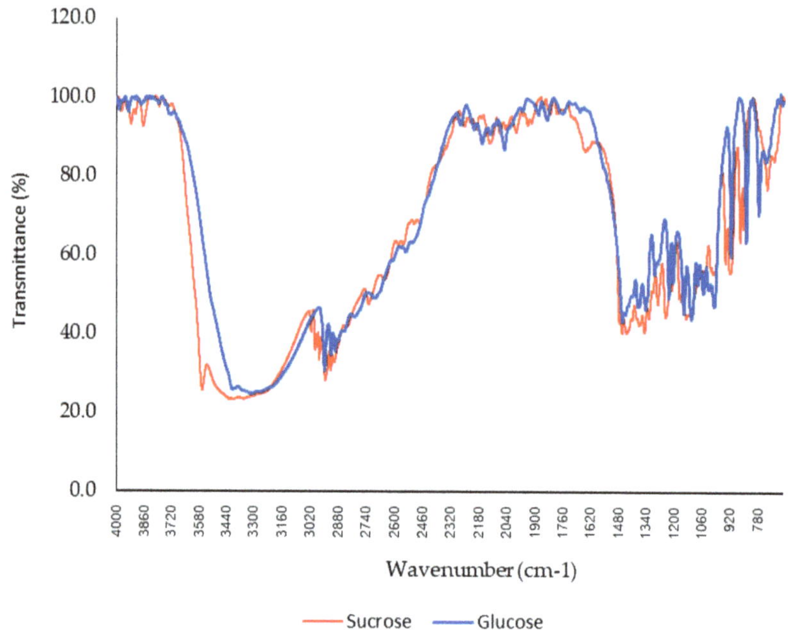

Figure 3.2. ATR-MIR spectra of pure sugar reference standards 10% (w/w) dissolved in d water. Source: [32] (unpublished data).

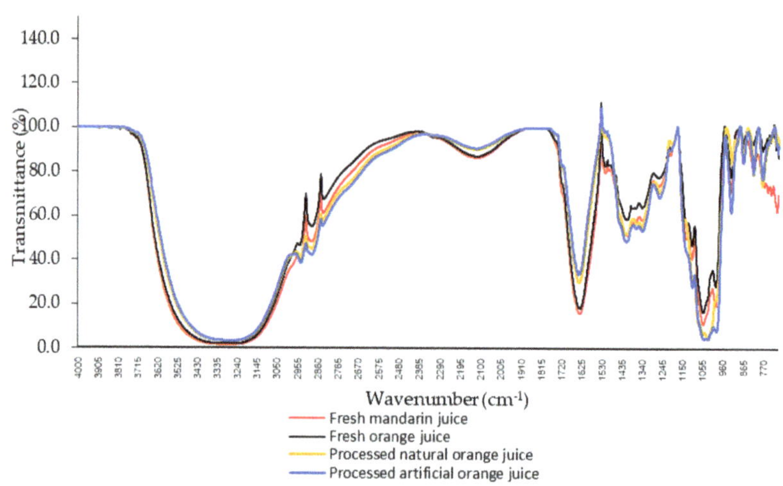

Figure 3.3. ATR-MIR spectra of fresh and processed orange juices. Source: [32] (unpublished data).

850–750 cm^{-1}, which can be assigned to their fingerprint region [33]. Moreover, processed orange juices were developed using natural ingredients, including fresh orange juice, acids, and sucrose (yellow line), and artificial ingredients, including sucrose and fructose syrup, acids, and artificial flavors (blue line). Their spectra also indicated their different vibration bonding [35].

3.8 Multivariate analysis of the FTIR results

Statistical analysis of analytical results has mostly been limited to univariate analysis, particularly with one-way analysis of variance (ANOVA) model and followed by post hoc tests, including Turkey 's range test or Duncan's test. However, high-throughput (HT) analysis, including the application of chromatographic and spectroscopic techniques, is becoming more important. Since it makes large-scale experiments and tests a large number of compounds, large data sets are generated in which essential information may not be indicated. Therefore, the application of multivariate statistics analysis for data exploration that involve many variables has been implemented and reported in the literature. Multivariate analysis of variances (MANOVA) or chemometrics are applied to maximize the information extracted from a data set and to optimize processes, recognize patterns, multivariate calibration, determine fraud and name origin, and classify products [36]. Chemometric techniques include unsupervised and supervised approaches. The unsupervised techniques such as principal components analysis (PCA) summarize variation in the data, regardless of the response, while the supervised techniques including partial least square (PLS) evaluate the variables to find a combination that explained the best causal relationships [37]. The large-scale information of the data matrix has led to the application of chemometric techniques, such as principal component analysis, discriminant analysis (PCA-DA) and partial least-squares regression. Therefore, the application of chemometrics for calibration, validation, and comparison of main components as determined by IR is indispensable [38]. The two most common multivariates analyses employed for the quantification of adulteration are PCA and PLS. PCA is undertaken to depict variability in a data set, such as an IR spectrum, using as few variables as possible. A linear combination of the original variables is then visualized in a score plot, where the position of each sample is positioned according to the values of the principal components (PCs) [36]. A PCA approach provides the visualization of the similarity and dissimilarity among suspected adulterated foods and authenic food compositions, with the expectation that the samples should have separated clusters obtaning from the scatter plot [38]. However, the supervised classification techniques, including linear discriminant analysis (LDA), k nearest neighbors (kNN), and partial least-squares discriminant analysis (PLS-DA), aim to individually assign unknown samples to a predefined class based on characteristics of samples [37]. These techniques all require at least two defined classes of samples and classify unknown samples to the class that it most closely resembles based on a set of characteristics. Recently, the classification methods using soft supervised learning models, such as Soft Independent Modeling Class Analogy (SIMCA), have been used to analyze suspected adulteration of foods. In addition, FTIR in conjunction with these multivariate analyses for the assessment of fruit juice authenticity has been studied and reported in the literatures [4].

3.9 Assessing the authenticity of fruit juices

The application of FTIR spectrometry combined with chemometric analysis of spectral data in the authentication of juice-to-juice adulteration has been reported in many studies. Table 3.2 shows a summary of studies of beverage adulteration and authenticity with fruit juices. In addition to juice-to-juice adulteration, FTIR spectrometry combined with chemometric analysis of spectral data was able to predict the adulteration of fruits juice by the addition of sweeteners (table 3.3). Meanwhile, an improper storage and transportation of non-alcoholic beverages, including fruit juices, over longer periods, can induce Maillard reaction, degrading nutritional components and generating genotoxic and carcinogenic by-products such as furfural and 5-hydroxymethylfurfural (HMF), rendering beverages unsafe for human consumption [52]. ATR approches can be undertaken to detect furfural and HMF in the same samples of fruit juice (table 3.4). In addition, sibutramine may be illicitly included in herbal slimming foods and supplements that are marketed as 100% natural to enhance weight loss. These compounds can be mixed in fruit juice to develop a fruit supplement [53]. Recently, Cassani *et al* [55] applied ATR-FTIR in the detection of fructooligosaccharide (FOS) addition in authentic juice (table 3.5). Furthermore, the determination of adulteration of fruit juice as a raw material in foodstuffs, including traditional pekmez-like high sugar containing products, using spectroscopic methods has been carried out (table 3.6).

3.10 FTIR in combination with other techniques

Several studies have reported the combination of FTIR analysis with other analytical chemical techniques, such as UV/Vis [57], NIR [58] or Raman spectroscopy [59], liquid chromatography coupled to either UV/Vis [57] or mass spectrometry (LC/MS) detectors [60], and DNA barcoding [61] to either compare a newly-developed FTIR method with validated or standardized methods or to complement the FTIR analysis.

3.11 Conclusions

Vibrational spectroscopic techniques, including FTIR spectroscopy, are emerging as powerful techniques to monitor the adulteration and authenticity of foods. Recently, governments, food industies, and consumers have demanded a rapid, non-destructive, and cheap method to detect and monitor new or unsuspected adulterants and contaminants, such as natural and artificial raw materials and/or chemical compounds. The assessment of adulteration and authenticity of foods requires the development and validation of analytical methods and their implementation. The application of IR combined with multivariates analysis or chemometrics provides a promising solution for the discrimination, classification, quantification, and identification of foods and beverages compositions. Therefore, FTIR spectroscopy will enable food manufacturers to rapidly evaluate the quality of their foods. Furthermore, the combination of the FTIR method with the other analytical techniques should be studied and developed to validate or standardize methods, or to complement the FTIR analysis.

Table 3.2. Summary of studies of beverage adulteration and authenticity with fruit juices.

Beverages	Adulterants	Short method description	Spectra range (cm^{-1})	Multivariate analysis	References
Pomegranate juice	Pomegranate juice concentrate (PJC) with grape juice concentrate (GJC).	An FTIR spectrophotometer coupled with ZnSe crystal was used over 4000–650 cm^{-1} region.	1780–1685 cm^{-1} region (C=O stretching) showed PJC and GJC dissimilarity.	PCA: (1) differentiate pure PJC and GJC samples and (2) classify adulterated (containing 2%–14% v/v GJC) and pure PJC samples.	[39]
Freshly squeezed orange Juice	100% concentrated orange juices at 0% to 30% (v/v).	ATR-FTIR was used to detect orange juice adulteration at 800–4000 cm^{-1} region.	1500–950 cm^{-1} (deformation of $–CH_2$ and deformation of C–C–H and H–C–O) indicated sugars and organic acids.	PCA could not discriminate adulteration levels between samples.	[40]
100% squeezed juices (pineapple, orange, and apple)	Grape juice at (5%, 10%, 15%, 20%, 30%, 40%, and 50%).	Spectra were recorded in the range of 952–3070 cm^{-1}.	900–1400 cm^{-1} (glucose, sucrose, and fructose) was presented. 000–1153 cm^{-1} (C–O and C–C stretching) and 1199–1400 cm^{-1} (O–C–H, C–C–H, and C–O–H bonds) was presented.	PCA and HCA results showed a clear clustering trend in the samples according to the type of juice analyzed.	[41]
Pomegranate juice	Dark colored sour cherry and black carrot juices.	ATR-FTIR spectrophotometer equipped with a horizontal ZnSe. and a doped triglycine sulfate (DTGS) detector. Measurement range was 4000–800 cm^{-1}.	1500–900 cm^{-1} (organic acids and sugars)3800–2790 cm^{-1} (–OH group of water and C–H stretching of organic acids)1300–1000 cm^{-1} (C–O stretching of organic acids) 1800–900 cm^{-1} (C–O and –OH groups of phenolic compounds).	PLS and OPLS regression models estimated adulteration percentages of pomegranate juice.	[42]

Grape juice classification	A classification of grape juices related to agricultural system (conventional and organic).	FTIR spectrometer coupled with a ZnSe crystal and helium/neon laser was used in a range between 951 and 1550 cm^{-1}.	1550–950 cm^{-1} (C–O, C–C, C–N, and C–H) for phenolic compounds, sugars, and organic acids1700–1000 cm^{-1} (C=C–C stretching of aromatic ring 1510–1450 cm^{-1} (OH-bending of phenol 1410–1310 cm^{-1} (C–H bending aromatic plane.		[36]
Commercial nectars of single flavors (grape, orange, peach, and passion fruit)	Diluted with syrup, apple, and cashew juices.	ATR-FTIR spectra were generated between 4000 and 650 cm^{-1}.	Intense band at 3600–3000 cm^{-1} and a sharp peak 1640 cm^{-1}, related to O–H absorption of water. 1500 and 950 cm^{-1} (as glucose and fructose and organic acids as citric and malic acids) 1700–1500 cm^{-1} (C–O stretching) of pectin and acids, with overlapped by water absorption band. 1200–900 cm^{-1}, (C–C and C–O stretching bonds) fingerprint region to sugars and organic acids.	PLS improved by iPLS-OPS variable selection showed the highest predictive capacity to quantify the main fruit contents.	[43]

Table 3.3. Summary of studies of beverage adulteration and authenticity with sweeteners.

Beverages	Adulterants	Short method description	Spectra range (cm^{-1})	Multivariate analysis	References
Mango juice	Adulterated with 0%–27% sucrose.	ATR-FTIR and chemometrics were used to evaluated in 4000–650 cm^{-1}.	1476–912 cm^{-1}. were predicted added sugar content with correlation coefficient of 0.99 and 0.98.	PLS and PCA were used to classify the samples with or without added sugar content.	[44]
Apple juice	Adulterated with inverted cane syrup, beet sucrose, high fructose corn syrup, and a synthetic solution of fructose, glucose, and sucrose.	ATR-FTIR have been used to detect adulteration of single strength apple juice samples.	1150 and 1470 cm^{-1} (C–C–H, C–O–H, and O–C–H bending)700 and 1150 cm^{-1} (C–O and C–C stretching) 1020–1060 cm^{-1} (O–H vibrations. All dominated by water and sugar absorptions.	PLS was used to discriminate between authentic and adulterated apple juice samples. An overall correct classification rate of 96.5% was achieved; 96% of the authentic juices and 97.8% of the adulterated samples.	[45]
Apple juice	Corn syrup was added as a potential sweetener.	ATR-FTIR spectrometer equipped with a deuterated DTGS detector was used to acquire all spectra at 4000 cm^{-1}–400 cm^{-1}.	3750–2750 cm^{-1} (O–H stretching vibrations of O–H) related to water 1750–1550 cm^{-1} containing 1636 cm^{-1} (C=O stretching) due to carbonyl group1200–900 cm^{-1} was fingerprint region for major carbohydrates.	SIMCA and PCA-LDA) performed equally well in discriminating all adulteration. Both methods can classify pure juice samples from adulterated samples (1% v/v).	[46]
Apple, peach, orange, pineapple,	Quantification of carbohydrates.	FTIR spectrometer equipped with a deuterated L-alanine	900–1400 cm^{-1}, the fingerprint region of glucose, fructose and	PCA was able to group of apple, orange, and peach juices.	[47]

and pear juices		doped triglycene sulfate (DLATGS) detector was used at 600–3500 cm^{-1}.	sucrose900–1153 cm^{-1} (C–O and C–C stretching)1199–1400 cm^{-1} (O–C–H, C–C–H and C–O–H bending vibrational).		
Coconut water	Sugar adulteration with binary sugars, mixed sugars, and high fructose corn syrup.	Infrared band region between 1200 and 950 cm^{-1}. FTIR Spectrometer equipped with a diamond crystal was used for spectral acquisition at 500–4000 cm^{-1}.	3700–2700 cm^{-1} (OH and C–H stretching)1700–1600 cm^{-1} (C=O stretching) of fructose and aldehyde (CH=O bending) of glucose 200 and 800 cm^{-1}, the fingerprint (C–C, C–O, O–C–H, and C–O–H), the functional groups.	PCA showed data clusters discernible by the type and level of substituted sugars, enabling visual diagnosis of the similarity and dissimilarity of level of substituted sugars. PLS regression indicated the prediction of adulterant sugars at single-digit percent substitutions.	[48]
Apple juice	Quantitative analysis of total sugar in processed fruit juices.	FTIR was used for spectral measurement. The juice samples were measured without any preparation directly on the KBr pellet.	885–1900 cm^{-1}, the sugar fingerprint region, was used for calibration and prediction.		[49]
Mango, pineapple, and grapes juices.	Quantification of sucrose content.	ATR-FTIR was used for direct to a detachable ZnSe ATR crystal.	2260–2100 cm^{-1} (the presence of alkyne group1650–1580 cm^{-1} (N–H bending) the presence of 1°-		[50]

(*Continued*)

Table 3.3. (*Continued*)

Beverages	Adulterants	Short method description	Spectra range (cm^{-1})	Multivariate analysis	References
			amines1250–1020 cm^{-1} (C–N stretch) the presence of aliphatic amines group 2750 and 3750 cm^{-1} indicated sucrose (1%–5%) with water absorbance.		
Pomegranate molasses syrups	Adulterated with date molasses.	FTIR spectrometer equipped with a multi-bounce ZnSe crystal was utilized as a qualitative control at 4000–500 cm^{-1} range.	3320 cm^{-1} (O–H stretching) 1720; 1640; 1230; 1060; and 910 cm^{-1} (C=O stretching, C=C– stretching, C–O stretching, and O–H stretching.		[51]

Table 3.4. Summary of studies of beverage adulteration and authenticity by natural or artificial compounds.

Beverages	Adulterants	Short method description	Spectra range (cm^{-1})	Multivariate analysis	References
Orange juice	Furfural and 5-hydroxymethylfurfural (HMF).	Furfural was characterized by ATR-FTIR spectroscopy in the range of 3200–750 cm^{-1}.	Specific to interaction product of (a) furfural (1654.09 cm^{-1}) and its derivative (b) 5-hydroxymethylfurfural (1662.86 cm^{-1}).		[52]
Tea and coffee	Sibutramine adulteration.	ATR-FTIR spectra of all samples were recorded at mid-infrared region within range of 4000–650 cm^{-1}.	Spectrum had significant vibrational bands at 3418, 2963, 2865, 2698, 1491, 1428, 1407, 1370, 1091, 1010, 833, and 822 cm^{-1}. These bands were associated with the chemical groups of components present in the chemical structure of sibutramine.	HCA and PCA are involved with arrangement of spectral data.	[53]
Grape-derived spirits	Predicting the alcoholic strength, methanol, acetaldehyde, and fusel alcohols content.	FTIR–ATR spectra of all samples were acquired in the wavelength region 4000–400 cm^{-1}.	1500 and 900 cm^{-1} (C–C and C–O vibrations) in volatile compounds of the beverages 1456 cm^{-1} (C–OH-bending deformation 3314 and 1646 cm^{-1}) that are assigned to water, to OH-stretching, and to OH-bending, respectively.	Quantitative estimation by applying PLS regression models. For methanol quantification, the spectral range between 1607 and 977 cm^{-1} was selected; for alcoholic strength between 3057–2864 + 2236–1920 + 1292–663 cm^{-1}.	[54]

Table 3.5. Summary of studies of the quantification of fructooligosaccharides using FTIR as an analysis technique.

Beverages	Adulterants	Short method description	Spectra range (cm^{-1})	Multivariate analysis	References
Strawberry juices	Fructooligosaccharides (FOS) are added to strawberry juices.	Quantifying FOS juices preserved for up to 14 days.	1134, 1034 cm^{-1} was identified the presence of FOS during storage.	PCA shows the different treatments. The groups observed explained more than 94% of the variance and were related to changes in the composition during storage.	[55]

Table 3.6. Summary of studies of the adulteration and authenticity of fruit juice as a raw material.

Beverages	Adulterants	Short method description	Spectra range (cm^{-1})	Multivariate analysis	References
Pekmez, a traditional Turkish food produced by concentration of fruit juice	Glucose syrup.	ATR-FTIR equipped with a single bounce diamond crystal and a deuterated triglycine sulfate detector were determined to be in the MIR range of 400–4000 cm^{-1}.	2908 cm^{-1} (C–H stretching) of carboxylic acid 800 and 1500 cm^{-1}. related to major sugars 1200–1500 cm^{-1} associated with deformation of –CH2, and angular deformation of C–C–H and H–C–O linkages. 950–1200 cm^{-1} associated to C–O and C–C stretching of carbohydrates.	PLS-DA was used to discriminate pure and adulterated grape, carob, and mulberry pekmez samples. After PLS-DA discrimination, the PLS technique was utilized to quantitatively determine the glucose syrup addition to various pekmez samples.	[56]

Remark:
FTIR = Fourier-transform infrared spectroscopy
ATR-FITR = Attenuated total reflectance Fourier-transform infrared spectroscopy
OPLS-DA = Orthogonal partial least-squares discriminant analysis
PCA = Principal component analysis
PLS-DA = Partial least-squares discriminant analysis
HCA = Hierarchical cluster analysis
PCA = Principle component analysis
SIMCA = Soft independent modeling by class analogy

References

[1] The 2023 World Population (https://worldpopulationreview.com) (accessed 10 March 2023)

[2] Revenue of the food industry worldwide 2014–2027 (https://statista.com/forecasts/1243605/revenue-food-market-worldwide) (accessed 10 March 2023)

[3] Acri G, Testagrossa B and Vermig G 2016 FT-NIR analysis of different garlic cultivar *J. Food Meas. Charact* **10** 127–36

[4] Dasenaki M E and Thomaidis N S 2019 Quality and authenticity control of fruit juices-a review *Molecules* **24** 1014

[5] Momtaz M, Bubli S Y and Khan M S 2023 Mechanisms and health aspects of food adulteration: a comprehensive review *Foods.* **12** 199

[6] Spink J and Moyer D C 2011 Defining the public health threat of food fraud *J. Food Sci.* **76** R157–63

[7] Robson K, Dean M, Haughey S and Elliott C 2021 A comprehensive review of food fraud terminologies and food fraud mitigation guides *Food Control* **120** 107516

[8] GAO 1995 *Fruit Juice Adulteration: Detection is Difficult, and Enhanced Efforts Would Be Costly* RCED-96-18 U.S. Government Accountability Office

[9] Kendall H, Clark B, Rhymer C, Kuznesof S, Hajslova J, Tomaniova M, Brereton P and Frewer L 2019 A systematic review of consumer perceptions of food fraud and authenticity: a European perspective *Trends Food Sci. Technol.* **94** 79–90

[10] Pappalardo L 2022 Pomegranate fruit juice adulteration with apple juice: detection by UV–visible spectroscopy combined with multivariate statistical analysis *Sci. Rep.* **12** 5151

[11] Maestrello V, Solovyev P, Bontempo L, Mannina L and Camin F 2022 Nuclear magnetic resonance spectroscopy in extra virgin olive oil authentication *Compr. Rev. Food Sci. Food Saf.* **21** 4056–75

[12] Keskin M, Arslan A, Soysal Y, Sekerli Y E and Celiktas N 2021 Feasibility of a chromameter and chemometric techniques to discriminate pure and mixed organic and conventional red pepper powders: a pilot study *J. Food Process. Preserv.* **46** e15846

[13] Teye E, Elliott C, Sam-Amoah L K and Mingle C 2019 Rapid and nondestructive fraud detection of palm oil adulteration with Sudan dyes using portable NIR spectroscopic techniques *Food Addit. Contam. Part* A **36** 1589–96

[14] Trifković J, Andrić F, Ristivojević P, Guzelmeric E and Yesilada E 2017 Analytical methods in tracing honey authenticity *J. AOAC Int.* **100** 827–39

[15] Muthukumaran P, Kartgikeyan R and Kumaravel S 2022 *A Comprehensive Guide To-Quality Analysis of Fruit Juices and Soft Drink—Analytical Procedures* (Thanjavur: Skyfox Publishing Group)

[16] Li J, Zhang C, Liu H, Liu J and Jiao Z 2020 Profile of sugar and organic acid of fruit juices: a comparative study and implication for authentication *J. Food Qual.* 7236534

[17] Sheikha A F E, Mokhtar N F K, Amie C, Lamasudin D U, Isa N M and Mustafa S 2017 Authentication technologies using DNA-based approaches for meats and halal meats determination *Food Biotechnol.* **31** 281–315

[18] Wu Y, Li M, Yang Y, Jiang L, Liu M, Wang B and Wang Y 2018 Authentication of small berry fruit in fruit products by DNA barcoding method *J. Food Sci.* **83** 1494–504

[19] Sobolev A P, Thomas F, Donarski J, Ingallina C, Circi S, Marincola F C, Capitani D and Mannina L 2019 Use of NMR applications to tackle future food fraud issues *Trends Food Sci. Technol.* **91** 347–53

[20] Psomiadis D, Zisi N, Koger C, Horvath B and Bodiselitsch B 2018 Sugar-specific carbon isotope ratio analysis of coconut waters for authentication purposes *J. Food Sci. Technol.* **55** 2994–3000
[21] Bureau S, Cozzolino D and Clark C J 2019 Contributions of Fourier-transform mid infrared (FT-MIR) spectroscopy to the study of fruit and vegetables: a review *Postharvest Biol. Technol.* **148** 1–14
[22] Wang L, Sun D-W, Pu H and Cheng J-H 2017 Quality analysis, classification, and authentication of liquid foods by near-infrared spectroscopy: a review of recent research developments *Crit. Rev. Food Sci. Nutr.* **57** 1524–38
[23] Włodarska K, Szulc J, Khmelinskii I, Sikorska E and E Non-destructive 2019 determination of strawberry fruit and juice quality parameters using ultraviolet, visible, and near-infrared spectroscopy *J. Sci. Food Agric* **99** 5953–61
[24] Valand R, Tanna S, Lawson G and Bengtström L 2020 A review of Fourier Transform Infrared (FTIR) spectroscopy used in food adulteration and authenticity investigations *Food Addit. Contam. Part* A **37** 19–38
[25] FT-IR Spectroscopy Attenuated Total Reflectance (ATR): Technical Note (https://perkinelmer.com) (accessed 17 April 2023)
[26] Arslan F N, Akin G, Elmas Ş N K, Üner B, Yilmaz I, Janssen H-G and Kenar A 2020 FT-IR spectroscopy with chemometrics for rapid detection of wheat four adulteration with barley four *J. Consum. Prot. Food Saf.* **15** 245–61
[27] Rodriguez-Saona L E and Allendorf M E 2011 Use of FTIR for rapid authentication and detection of adulteration of food *Annu. Rev. Food Sci. Technol.* **2** 467–83
[28] Jamwal R, Kumari A S, Sharma S, Kelly S and Cannavan A 2021 Recent trends in the use of FTIR spectroscopy integrated with chemometrics for the detection of edible oil adulteration *Vib. Spectrosc.* **113** 103222
[29] Coates J 2000 Interpretation of infrared spectra: a practical approach In *Encyclopedia of Analytical Chemistry* ed R A Meyers (Chichester: Wiley) pp 10815–37
[30] Park Y-S, Im M H, Ham K-S, Kang S-G, Park Y-K, Namiesnik J, Leontowicz H, Leontowicz M, Trakhtenberg S and Gorinstein S 2015 Quantitative assessment of the main antioxidant compounds, antioxidant activities and FTIR spectra from commonly consumed fruits, compared to standard kiwi fruit *LWT-Food Sci. Technol.* **63** 46–352
[31] Shurvell H F 2002 Spectra-structure correlations in the mid and far infrared In *Handbook of Vibrational Spectroscopy* ed J M Chalmers and P R Griffiths (Chichester: Wiley) pp 1783–816
[32] Hamidom M and Wongsa P 2015 Authenticity assessment of natural and artificial compounds in orange juice *Program in Food technology, School of Agro-Industry, Mae Fah Luang University* (unpublished data)
[33] Bureaua S, Cozzolinob D and Clarkc C J 2019 Contributions of Fourier-transform mid infrared (FT-MIR) spectroscopy to the study of fruit and vegetables: a review *Postharvest Biol. Technol.* **148** 1–14
[34] Ellis D I, Brewster V L, Dunn W B, Allwood J W, Golovanovc A P and Goodacre R 2012 Fingerprinting food: current technologies for the detection of food adulteration and contamination *Chem. Soc. Rev.* **41** 5706–27
[35] Leopold L F, Leopold N, Diehl H-A and Socaciu C 2011 Quantification of carbohydrates in fruit juices using FTIR spectroscopy and multivariate analysis *J. Spectrosc.* **26** 93–104

[36] Junges C H, Guerra C C, Gomes A A and Ferr˜ao M F 2022 Green analytical methodology for grape juice classification using FTIR spectroscopy combined with chemometrics *Talanta Open* **6** 100168
[37] Scott E R and Crone E E 2021 Using the right tool for the job: the difference between unsupervised and supervised analyses of multivariate ecological data *Oecologia* **196** 13–25
[38] Valand R, Tanna S, Lawson G and Bengtström L 2020 A review of Fourier Transform Inrared (FTIR) spectroscopy used in food adulteration and authenticity investigations *Food Addit. Contam. Part* A **37** 19–38
[39] Vardin H, Tay A, Ozen B and Mauer L 2008 Authentication of pomegranate juice concentrate using FTIR spectroscopy and chemometrics *Food Chem.* **108** 742–8
[40] Shen F, Wu Q, Su A, Tang P, Shao X and Liu B 2014 Detection of adulteration in freshly squeezed orange juice by electronic nose and infrared spectroscopy *Czech J. Food Sci.* **34** 224–32
[41] Calle J L P, Ferreiro-González M, Ruiz-Rodríguez A, Fernández D and Palma M 2022 Detection of adulterations in fruit juices using machine learning methods over FT-IR spectroscopic data *Agron.* **12** 683
[42] Aykac B, Cavdaroglu C and Ozen B 2023 Authentication of pomegranate juice in binary and ternary mixtures with spectroscopic methods *J. Food Compos. Anal.* **117** 105100
[43] Sheng C, Miaw W, Assis C, Silva A R C S, Cunha M L, Sena M M and de Souza S V C Determination of main fruits in adulterated nectars by ATR-FTIR spectroscopy combined with multivariate calibration and variable selection methods *Food Chem.* **254** 2018 – 272–80
[44] Jha S N and Gunasekaran S 2010 Authentication of sweetness of mango juice using Fourier transform infrared-attenuated total reflection spectroscopy *J. Food Eng.* **101** 337–42
[45] Kelly J F D and Downey G 2005 Detection of sugar adulterants in apple juice using Fourier Transform Infrared spectroscopy and chemometrics *J. Agric. Food Chem.* **53** 3281–6
[46] A S *et al* 2020 Qualitative and quantitative evaluation of corn syrup as a potential added sweetener in apple fruit juices using mid-infrared spectroscopy assisted chemometric modeling *LWT-Food Sci. Technol.* **131** 109749
[47] Leopold L F, Leopold N, Diehl H-A and Socaciu C 2011 Quantification of carbohydrates in fruit juices using FTIR spectroscopy and multivariate analysis *J. Spectrosc* **26** 93–104
[48] Teklemariam T A, Moisey J and Gotera J 2021 Attenuated Total Reflectance-Fourier transform infrared spectroscopy coupled with chemometrics for the rapid detection of coconut water adulteration *Food Chem.* **355** 129616
[49] Singh S and Singh S 2016 Sugar analysis in apple juice by FTIR spectroscopy with PLS regression *Int. J. Innov. Sci. Eng.* **2**
[50] Nair R, Venkatesh S, Athmaselvi K A and Thakur S 2015 Rapid estimation and quantification of sucrose content in fruit juices using Fourier transform infrared–attenuated total reflectance (FTIR–ATR) spectroscopy *Food Measure* **10** 24–31
[51] Darra N E, Rajha H N, Al-Oweini F S R, Maroun R G and Louka N 2017 Food fraud detection in commercial pomegranate molasses syrups by UV–VIS spectroscopy, ATR-FTIR spectroscopy and HPLC methods *Food Control* **78** 132–7
[52] Mehrotra S, Rai P and Sharma S K 2022 A quick and simple paper-based method for detection of furfural and 5-hydroxymethylfurfural in beverages and fruit juices *Food Chem.* **377** 131532

[53] Cebi N, Yilmaz M T and Sagdic O 2017 A rapid ATR-FTIR spectroscopic method for detection of sibutramine adulteration in tea and coffee based on hierarchical cluster and principal component analyses *Food Chem.* **229** 517–26

[54] Anjos O, Santos A J A, Estevinho L M and Caldeira I 2016 FTIR–ATR spectroscopy applied to quality control of grape-derived spirits *Food Chem.* **205** 28–35

[55] Cassani L, Santos M, Gerbino E, del Rosario Moreira M and Gómez-Zavaglia A 2018 A combined approach of infrared spectroscopy and multivariate analysis for the simultaneous determination of sugars and fructans in strawberry juices during storage *J. Food Sci.* **83** 631–8

[56] Yaman N, Velioglu S D and S 2019 Use of attenuated total reflectance—Fourier transform infrared (ATR-FTIR) spectroscopy in combination with multivariate methods for the rapid determination of the adulteration of grape, carob and mulberry PEKmez *Foods* **8** 231

[57] Aboulwafa M M, Youssef F S, Gad H A, Sarker S D, Nahar L, Al-Azizi M M and Ashour M L 2019 Authentication and discrimination of green tea samples using UV–vis, FTIR and HPLC techniques coupled with chemometrics analysis *J. Pharm. Biomed. Anal.* **164** 653–8

[58] Huang F, Song H, Guo L, Guang P, Yang X, Li L, Zhao H and Yang M 2020 Detection of adulteration in Chinese honey using NIR and ATR-FTIR spectral data fusion *Spectrochim. Acta A Mol. Biomol. Spectrosc.* **235** 118297

[59] Cebi N, Arici M and Sagdic O 2021 The famous Turkish rose essential oil: characterization and authenticity monitoring by FTIR, Raman and GC–MS techniques combined with chemometrics *Food Chem.* **354** 129495

[60] Widyanti H P, Vevi M, Ida M, Syamsul H, Nandi S and Muchtaridi M 2022 Possibilities of liquid chromatography mass spectrometry (LC-MS)-based metabolomics and lipidomics in the authentication of meat products: a mini review *Food Sci. Anim. Resour* **42** 744–61

[61] Naim N, Ennahli N, Hanine H, Lahlali R, Tahiri A, Fauconnier M-L, Madani I and Ennahli S 2022 ATR-FTIR spectroscopy combined with DNA barcoding and GC–MS to assess the quality and purity of saffron (*Crocus sativus* L.) *Vib. Spectrosc.* **123** 103446

IOP Publishing

Advances in Spectroscopic Analysis of Food and Drink

Ashutosh Kumar Shukla

Chapter 4

Comparison between binary and multiclass partial least square with discriminant analysis: a case study on coffee classification based on near-infrared spectra

Fabiana de Carvalho Pires, Rosemary Gualberto Fonseca Alvarenga Pereira, Michel Rocha Baqueta, Patrícia Valderrama and Roney Alves da Rocha

Chemometric methods have become essential for solving classification problems in many scientific areas. Their application in the evaluation of coffee roasting profiles is an important issue due to the large range of consumer preferences. A typical example in this classification field is partial least square with discriminant analysis (PLS-DA). Several studies have applied binary PLS-DA to solve multiclass problems, while the existing multiclass version of the method is neglected. In this study, near-infrared (NIR) spectroscopy has been used to classify coffee beans and ground coffees concerning their roasting degrees (light, class 1; medium, class 2; or dark, class 3) using two approaches: binary PLS-DA, or 'one against all', and multiclass PLS-DA. The results of quality classification show that sensitivity and specificity vary insignificantly with the PLS-DA versions. When one target class is discriminated against the other available classes in binary PLS-DA, the results are very close to those results obtained in multiclass discrimination. Models for ground coffees and coffee beans with light and dark roasting degrees were better classified in prediction sets, while models for medium roasting degrees were poorly classified due to intermediate characteristics. Multiclass PLS-DA is more appropriate for implementation in industrial applications.

4.1 Introduction

The quality of coffee is dictated by many factors, such as production, harvesting, processing, drying, storage, and the roasting process (Borém *et al* 2016, Pereira *et al* 2020). Among these steps, the roasting process plays an important role in the

doi:10.1088/978-0-7503-5573-5ch4 © IOP Publishing Ltd 2023

formation of the aroma, flavor, and color characteristics of coffee (Craig *et al* 2018, Belchior *et al* 2019). The way in which roasting is conducted (time and temperature) allows the development of different sensory attributes and cup quality (Poisson *et al* 2017, Debona *et al* 2021).

The evaluation of the roasting degree in coffee is important in several aspects—either to monitor and interrupt the process, or to ensure the final roast profile of the coffee. In the industry, light, medium, and dark roasting are the main degrees for coffee marketing. Several parameters can be considered to evaluate the degree of roasting in coffee, such as color, aroma, volume, and the temperature of coffee beans. However, the most frequently used is based on visual inspection, which is not always the best method because it suffers from subjectivity (Hernández *et al* 2008a, 2008b). Another possibility that is quite common in the industry is the Agtron roast color classification system (or the Agtron method), which is a colorimetric technique that measures the amount of energy reflected by the surface of beans or ground roasted coffee (Agtron and SCAA 2010, Agtron Inc. 2010, Nansen *et al* 2016, Santos *et al* 2016).

Besides being useful for the operator in the industry, the roasting degree measured by the Agtron method can be explored in other ways, such as for the development of complementary tools for monitoring the roasting in the final product. An example is the use of infrared spectroscopy (Barbin *et al* 2014, Fioresi *et al* 2021). In the scientific literature, near-infrared (NIR) spectroscopy with chemometric tools has been reported as an alternative and versatile method to monitor the roasting process in real-time (Santos *et al* 2016, Catelani *et al* 2018) and to measure the final degree of roasting in coffee (Alessandrini *et al* 2008, Bertone *et al* 2016, Yergenson and Aston, 2020, Pires *et al* 2021). Although NIR technology is mature in the field of coffee science and roast degree evaluation, its application for industrial purposes is still scarce and depends on proper chemometric treatment to build reliable models.

Chemometric methods have become essential for solving classification problems in several scientific areas. Their application to evaluate the degree of coffee roasting is an important issue due to differences in the consumer preferences (Marquetti *et al* 2016, Craig *et al* 2018, Belchior *et al* 2019, Giraudo *et al* 2019). PLS-DA is a typical example in this classification field. Although several studies have applied binary PLS-DA to solve multiclass problems, the existing multiclass version of the method has been neglected (Pomerantsev and Rodionova 2018). Multiclass PLS-DA has recently been used to discriminate and classify roasted coffee blends based on their cup quality and promising results have been presented for its implementation on an industrial scale (Baqueta *et al* 2021).

For industry, multiclass models may be easier for operations. In this study, NIR spectroscopy has been used to classify coffee beans and ground coffees concerning their roasting degrees (light, class 1; medium, class 2; or dark, class 3) in two approaches: binary PLS-DA (or 'one against all') and multiclass PLS-DA. The roasting degree drastically alters the characteristics of the coffee. For consumers who are not attentive, the degree of roasting may not be easily distinguishable. Meanwhile, the development of models to classify the profile of coffee roasting can be economically favorable for the industry because it expands the portfolio of coffee profiles and meets the different preferences of consumers.

4.2 Material and methods

4.2.1 *Coffee samples*

Green *Coffea arabica* L. (Catuaí Vermelho IAC 99) was harvested in 2018 at Três Porteiras farm in the state of Minas Gerais, Brazil. The green beans were dry processed (natural coffee) and had a sensory score of 81.5 according to cupping (SCA 2003).

4.2.2 *Roasting and grinding processes*

The roasting was carried out for non-defects and the coffee beans that were retained on sieves 16 and higher (17 and 18/64 inches; Pinhalense Inc., Espírito Santo do Pinhal, Brazil) (Borém *et al* 2016). Green coffee beans (100 g per sample) were roasted using a Probat TP 2 (Probat Leogap Inc., Curitiba, Brazil) with an initial temperature of 150 °C and maximum duration of 12 min (more details about roasting degree can be seen in table 4.1S, which is annexed in the supplementary material). To stop the thermal process, the samples were immediately transferred to the roaster-cooling tray.

Three roasting degrees were carried out on the green beans: light, medium, and dark. Each roasted coffee was evaluated during the process using Agtron disks (Agtron and SCAA 2010). The roasting process was stopped when the color's endpoint of the roasting was similar to disks Agtron disks between #75 and #45. After cooling the roasted coffees, a test was carried out using the Agtron spectrophotometer to certify that the roasted coffees were in the desired roast degree, according Pires *et al* (2021).

The roasted samples were split into two groups: coffee beans and ground coffees. The coffees were ground using a Guatemala K32S21 (Mahlkönig Inc., Hamburg, Germany) with a particle size of 20 Tyler mesh (SCA 2003).

4.2.3 *Agtron method*

Beans and ground roasted coffee samples were evaluated with the Agtron Roast Color Classification system to determine their roasting degrees (light, medium, or dark) and to ensure they were within the appropriate range. The Agtron value between #75 and #65 is light, between #65 and #55 is medium, and between #55 and #45 is dark (Agtron and SCAA 2010). An Agtron spectrophotometer M-basic II (Agtron Inc., Reno, United States of America) with scale ranges between 0.0 ± 5.0 and 133.0 ± 5.0 was used to measure exactly the Agtron value of each roast degree. Coffee samples were evaluated in quadruplicate and mean values were considered as the final roasting degree. Before analysis, the spectrophotometer was calibrated according to the manufacturer's guidelines (Agtron Inc. 2010).

Table 4.1. Sensitivity and specificity results obtained in the binary models.

Model	Training set		Cross-validation set		Prediction set	
	Sensitivity	Specificity	Sensitivity	Specificity	Sensitivity	Specificity
Models for bean samples						
Class 1 Light roasting degree	0.985	0.964	0.582	0.741	0.767	0.817
Class 2 Medium roasting degree	0.912	0.933	0.426	0.622	0.467	0.717
Class 3 Dark roasting degree	1.000	0.978	0.833	0.948	0.900	0.983
Models for ground samples						
Class 1 Light roasting degree	1.000	1.000	0.857	0.869	0.800	0.917
Class 2 Medium roasting degree	0.986	0.979	0.529	0.679	0.800	0.583
Class 3 Dark roasting degree	1.000	1.000	0.971	0.971	0.833	0.983

4.2.4 *NIR spectroscopy*

A Bruker Tensor 27 (Ettlingen, Germany) NIR spectrometer was used for spectral data collection. The settings were: a resolution of 4 cm^{-1} with 64 scans per sample in the region from 4000 to 10 000 cm^{-1} and 3111 variables per spectrum. The spectra were obtained at 45° with zinc selenide crystal. The spectra were recorded in beans and ground coffees samples. Ten different aliquots of samples were prepared, and the spectra of each aliquot was recorded into the roast degree (100 spectroscopic data for light degree, 100 for medium degree and 100 for dark degree), totaling 300 spectra from beans coffees and 300 from ground coffees.

4.2.5 *Data handling and preprocessing*

All data preprocessing and chemometric analysis were performed through Matlab software version R2007b (The MathWorks Inc., Natick, USA) and PLS-Toolbox

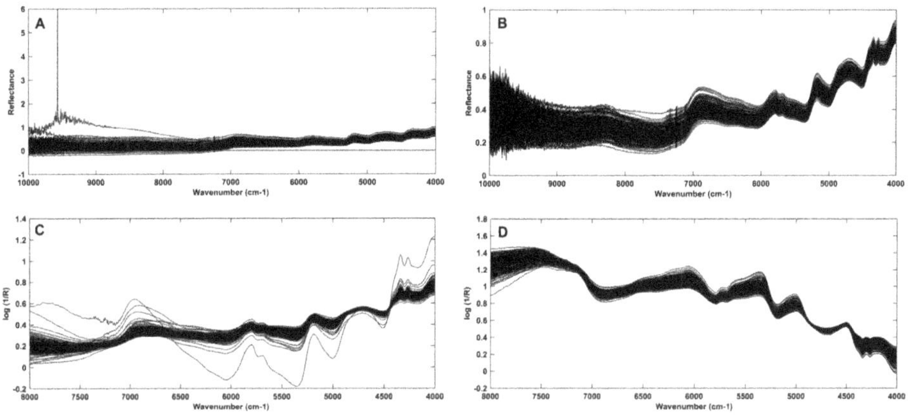

Figure 4.1. NIR spectra obtained directly in coffee samples. Raw spectra of beans (A) and ground (B) coffee samples. Preprocessed spectra of beans (C) and ground (D) coffee samples.

computational package version 5.2 (Eigenvector Technologies, Manson, Washington). Preprocessing techniques were applied to the original NIR spectra data sets to compensate for any changes in experimental conditions and to improve the results. The preprocessing techniques in both NIR spectra (from the coffee beans and ground coffees) were similar to make the results more comparable.

The original spectroscopic profiles obtained directly in the samples for both beans (n = 300) and ground (n = 300) are shown in figures 4.1(A) and (B), respectively. Two samples of beans coffee showed an unusual behavior before any pretreatment and were removed, resulting in 298 samples for modeling. In addition, due to the intense noise, the spectral region between 10 000 and 8000 cm^{-1} was excluded in both beans and ground spectra of the coffee samples. The remaining reflectance spectra were then transformed into absorbance. The pretreatments on the spectra included the multiplicative scatter correction (MSC) (Geladi *et al* 1985) and smoothing with the first-order polynomial applying the Savitzky–Golay algorithm with a window of 16 points (Golay 1964).

4.2.6 *Implementation of PLS-DA models*

The PLS-DA model consists of the matrix X ($N \times J$) that represents the instrumental response (NIR data sets), where N is the number of samples and J is the variable number. The vector Y corresponds to the sample class.

In this study, the class of each sample was established according to the degree of roasting (light, medium, and dark) based on the Agtron method (see figures 4.2 and 4.1S in the supplementary material). Binary models were built for the classification of each coffee class, three models for the classification of coffee beans and three models for the classification of ground coffees. Therefore, six binary models were built to classify each coffee class or target class. Furthermore, we investigated the potential of multiclass PLS-DA models for the classification of the samples, resulting in two models: one for coffee beans and the other for ground coffees. This test was

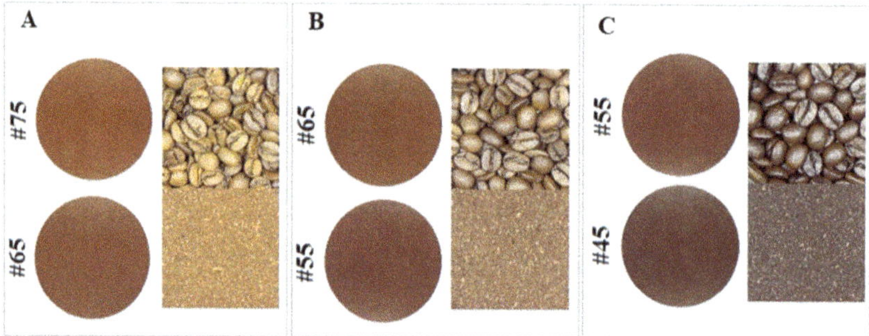

Figure 4.2. Classification of coffee roasting degree using Agtron Roast Color Classification System: A, moderately light to medium light (#75 to #65); B, medium light to medium (#65 to #55); and C, medium to moderately dark (#55 to #45).

carried out to check the efficiency of multiclass PLS-DA for industrial routine purposes, which would be simpler for coffee operators.

For the building and validation of the PLS-DA models, 70% of the samples were randomly selected for the training set and the remaining samples (30%) were used as the prediction set (test set). This was done for each class. The training samples were selected with the Kennard–Stone algorithm (Kennard and Stone 1969), which ensured the representativeness of the models. The samples in the prediction set had not been used to develop the classification models. The models were built using the cross-validation method, with contiguous blocks with 10 data splits. The optimal number of latent variables (LVs) was chosen in each model from a combined analysis of different parameters, including variance explained in Y block, class border evaluation by Bayesian decision (which minimizes classification errors), and receiver operating characteristics (ROC) plots (Ferreira 2015).

A step of outlier detection and elimination was also executed to improve the model's accuracy, remove samples with extreme values (which exhibited increased influence on the model), and to eliminate unmodelled residuals in the data responses. Possible outliers on the training and prediction sets also were evaluated using a test of leverage and Q residuals. While leverage represents how distant one sample is from the center of the data, Q residuals represent the unmodelled residuals. Samples can be considered to be outliers when they show simultaneously high leverage and high Q residuals (Alves and Valderrama 2015, Valderrama and Valderrama 2016, Dias *et al* 2018).

4.2.7 *Model quality evaluation*

The accuracy of the models was evaluated using the combined values of root mean squares error of cross-validation (RMSECV), root mean squares error of calibration (RMSEC), and root mean squares error of prediction (RMSEP) (Alves and Valderrama 2015, Valderrama and Valderrama 2016, Dias *et al* 2018).

Many parameters may be used to evaluate the efficiency of chemometric classification methods (López *et al* 2015, Valderrama and Valderrama 2016, Craig *et al* 2018, Dias *et al* 2018), and sensitivity and specificity values are most common (Alves and Valderrama 2015, Valderrama and Valderrama 2016, Dias *et al* 2018). Sensitivity (equation (4.1)) and specificity (equation (4.2)) rates can assume results from 0 (0%) to 1 (100%) and were calculated referring to the training set, to the internal cross-validation set, and to the prediction set considering true positive (TP) and negative (TN), and false-positive (FP), and negative (FN) rates (López *et al* 2015, de Santana *et al* 2016).

$$\text{Sensitivity} = \frac{N(\text{true positives})}{N(\text{true positives}) + N(\text{false negatives})} \tag{4.1}$$

$$\text{Specificity} = \frac{N(\text{true negatives})}{N(\text{true negatives}) + N(\text{false negatives})} \tag{4.2}$$

To assess the main regions of the spectrum responsible for the modeling of each class, the variable importance in projection (VIP) scores were calculated. A VIP score greater than 1 is considered important for the modeling, while VIP scores lower than 1 are less important (de Santana *et al* 2016).

4.3 Results and discussions

4.3.1 *Descriptive analyzes*

Details of Agtron scale are shown in table 4.1S. Bean coffees presented Agtron values (mean ± standard deviation) of #71.7±2.3 for light roasted samples, #62.2 ±3.4 for medium roasted, and #48.5±2.9 for dark roasted. Ground coffees samples presented Agtron values of #76.4±3.5 for light roasted samples, of #62.1±4.9 for medium roasted, and of #43.2±3.3 for dark roasted.

4.3.2 *Visual spectra evaluation*

Preprocessed NIR spectra of beans and ground coffees are shown, respectively, in figures 4.1(C) and (D). From these figures it can be seen that it is not possible to visually distinguish them based on their roasting degrees (light, medium, or dark roast).

Due to the lack of selectivity in NIR spectroscopy, it is difficult to conclude the specific compounds present in coffee samples simply by observing the spectra. However, based on the literature, some assignments of coffee compounds can be made. For example, vibrations in the NIR region at the 4337–4325 cm^{-1} can be attributed to sucrose and other carbohydrates (Ribeiro *et al* 2011). In the region between 4566 and 4562 cm^{-1}, proteins and chlorogenic acids present vibration modes (Ribeiro *et al* 2011). Finally, from 5862 to 5834 and from 2480 to 2488 cm^{-1} the vibrations can be attributed to lipids (Ribeiro *et al* 2011). In addition, the NIR spectra of roasted coffee are strongly affected by the water band intensity at 6896 cm^{-1} (first overtone of O-H stretching) and 5154 cm^{-1} (combination band O–H stretching and

O–H deformation). This occurs when the coffees are roasted darker due to the reduction of water (Esteban-Díez *et al* 2004, Alessandrini *et al* 2008).

Other vibrations can also be attributed in coffee spectroscopic profiles, such as for proteins at 6623 cm $^{-1}$ (the first overtone of N–H stretching) and chlorogenic acids at 5935 cm^{-1} (first overtone of C–H vibration in aromatic structures), lipids at 4669 cm^{-1}, quinic acid at 4771–4840 cm^{-1}, and caffeine at 4655 and 8666 cm^{-1} (Esteban-Díez *et al* 2004) In addition, others spectral regions (second overtone region of C=O, combination band and first overtone region of C=O and O–H bonds) are associated with organic acids, proteins, and carbohydrate vibrations (Barbin *et al* 2014).

4.3.3 *Optimization by outlier elimination*

During modeling, different symbols and colors were used to more easily differentiate BINARY from multiclass strategies. The results are discussed in detail in this section.

The first step in the development of PLS-DA models was to investigate the presence of possible outliers in data sets using a test based on leverage and Q residuals. Figure 4.2S in the supplementary material shows the results for binary models for beans and ground coffees), while figure 4.4(A) (beans coffee) and figure 4.5(A) (ground coffees) in the current text show the results for multiclass models. Some coffee bean samples in binary model (figures 4.2S(A)–(C)) and others in the multiclass model for coffee beans (figure 4.4(A)) simultaneously presented high leverage and high Q residuals, which is indicative of potential outliers. Therefore, these samples were removed and the models for the classification of beans coffees were rebuilt.

4.3.4 *Classification strategies*

PLS-DA can be used to confirm if a set of data belongs to the pre-established classes, and therefore it is necessary to have initial knowledge of the classes (Barker and Rayens 2003, Brereton and Lloyd 2014). In the PLS-DA classification, when the result of the sample is above the threshold value, it will be considered to belong to the modeled class. When the value is below the threshold, the sample does not belong to the modeled class (Marquetti *et al* 2016, Valderrama and Valderrama, 2016, de Santana *et al* 2020).

Figure 4.3S (for binary model) and figure 4.5S (for multiclass model) show the ROC curves. Table 4.2S (for binary model) and figure 4.3S (for multiclass model) show the optimal number of LVs, explained variance presented in Y block for each class modeled, and their respective classification errors. The combined analysis of these results indicates that the number of optimal LVs was correctly chosen to obtain the best results in terms of sensitivity and specificity. Coffee is a complex chemical matrix, which is full of variations and has been a challenge in chemometric modeling in many cases. Belchior *et al* (2016), Marquetti *et al* (2016), Craig *et al* (2018), Dias *et al* (2018), and Belchior *et al* (2019) achieved an explained variance in the Y block between 70% and 95% in previous coffee studies. Therefore, the results described in the present study are in this line.

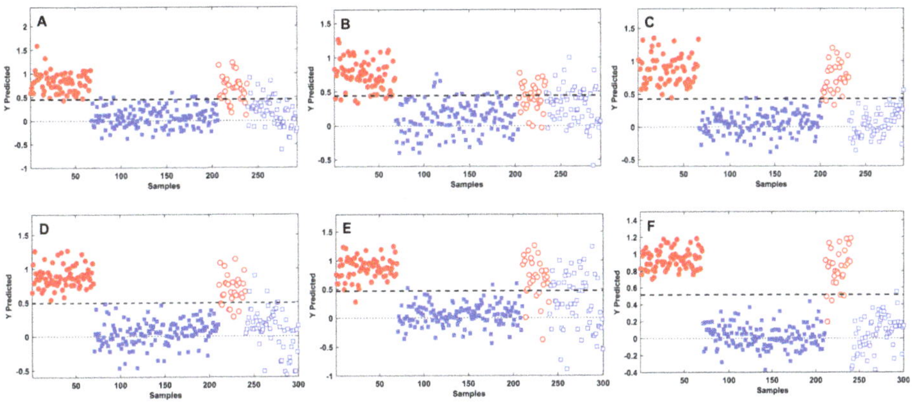

Figure 4.3. Estimated class values for training and prediction sets in the binary models for beans (A, class 1: light roasting degree; B, class 2: medium roasting degree; C, class 3: dark roasting degree) and ground (D, class 1: light roasting degree; E, class 2: medium roasting degree; F, class 3: dark roasting degree) coffees. Sample sets of training (●) and prediction (○) of the class of interest. Sets of training (■) and prediction (□) of the rest of the samples.

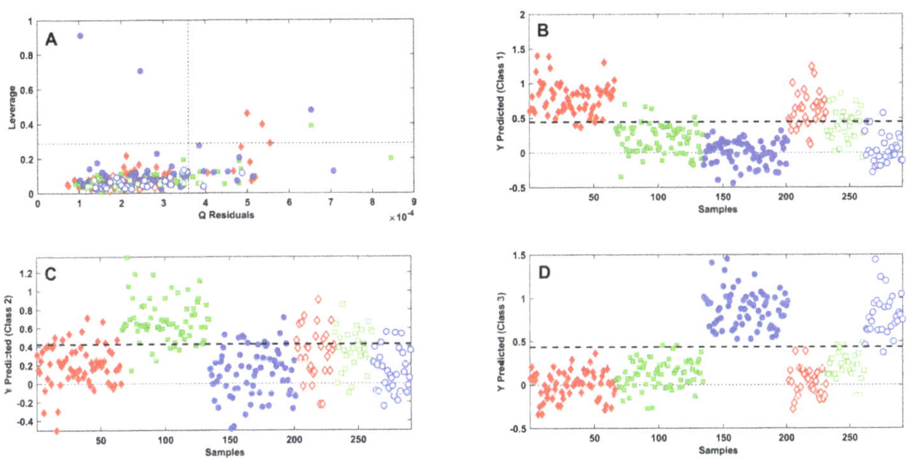

Figure 4.4. Outlier identification (A) and estimated class values for training and prediction sets in the multiclass model for the classification of beans (B, class 1: light roasting degree; C, class 2: medium roasting degree; D, class 3: dark roasting degree) coffees. Samples of training (◆) and prediction (◇) for class 1, light roasting degree. Samples of training (■) and prediction (□) for class 2, medium roasting degree. Samples of training (●) and prediction (○) for class 3, dark roasting degree.

RMSEC, RMSECV, and RMSEP values are shown in table 4.2 S (for binary model) and table 4.3S (for the multiclass model). Since these values are close for each class (light, medium, and dark roasting degree) and modeled a good agreement between them, this indicates that future classifications are accurate (Valderrama and Valderrama 2016, Dias *et al* 2018).

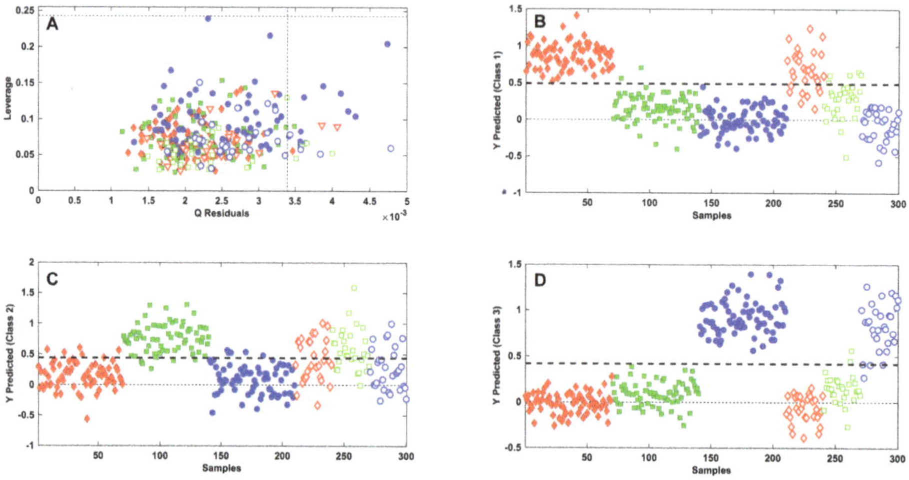

Figure 4.5. Outlier identification (A) and estimated class values for training and prediction sets in the multiclass model for the classification of ground coffees (B, class 1: light roasting degree; C, class 2: medium roasting degree; D, class 3: dark roasting degree) coffees. Samples of training (♦) and prediction (◇) for class 1, light roasting degree. Samples of training (■) and prediction (□) for class 2, medium roasting degree. Samples of training (●) and prediction (○) for class 3, dark roasting degree.

Figures 4.3 (A–E), 4.4 (B–D), and 4.5 (B–D) show the estimated class values for training and prediction sets in the PLS-DA models. Sensitivity and specificity were estimated through these plots. Sensitivity is the model's ability to correctly classify the samples when these samples are right. Meanwhile, specificity consists of an incorrect prediction of samples of other classes in a particular class. Thus, if the model does not present error in predicting a sample, this model presents specificity equal to 1 (Marquetti *et al* 2016, Valderrama and Valderrama 2016, Dias *et al* 2018, Giraudo *et al* 2019).

Table 4.1 (for the binary model) and table 4.2 (for the multiclass model) show that the final sensitivity and specificity for classifying, respectively, beans and ground coffee samples based on their roasting degree binary model achieved slightly better classifications than the multiclass model for both beans and ground coffees. The sensitivity and specificity ranged from 0.865 to 1.000 in the training sets considering beans and ground coffees samples in the three roasting degrees and in both PLS-DA strategies, which is considered satisfactory. Meanwhile, beans and ground coffees with light and dark roasting degree were better classified in prediction sets in both classification strategies than for medium roasting degree. The obtained sensitivity values multiclass models were 0.733 for light roasting degree ground coffees and 0.867 for dark roasting degree ground coffees, and the specificity values were 0.917 (light roasting degree) and 0.967 (dark roasting degree) For medium roasting degree beans, the multiclass model sensitivity values were 0.433 (coffee beans) and 0.667 (ground coffees), and the specificity values were 0.767 (beans) and 0.650 (ground),

Table 4.2. Sensitivity and specificity results obtained in the multiclass models.

Model	Training set		Cross-validation set		Prediction set	
	Sensitivity	Specificity	Sensitivity	Specificity	Sensitivity	Specificity
Models for bean samples						
Class 1 Light roasting degree	0.970	0.963	0.621	0.770	0.800	0.750
Class 2 Medium roasting degree	0.897	0.865	0.485	0.549	0.433	0.767
Class 3 Dark roasting degree	1.000	0.985	0.881	0.948	0.967	0.950
Models for ground samples						
Class 1 Light roasting degree	1.000	0.986	0.871	0.857	0.733	0.917
Class 2 Medium roasting degree	0.929	0.929	0.571	0.693	0.667	0.650
Class 3 Dark roasting degree	1.000	1.000	0.943	0.936	0.867	0.967

both samples were poorly classified. Through these results, it is possible to verify that in the prediction of new samples, the models work better for the classification of light or dark roasted coffees and that the medium roasting suffers from ambiguity in the models. To try to improve the performance of the model for classification of samples with medium roasting, the inclusion of more samples to increase the representativity could provide better results in the future. From an industrial perspective, a multiclass model would be an easier alternative for operators to classify mainly beans and ground coffees with light or dark roasting degree because it is multiclass, even with its slightly worse results.

Overall, the sensibility and specificity achieved in this study are comparable to previous studies of coffee, both using portable and benchtop NIR spectrometers (Santos *et al* 2012, Marquetti *et al* 2016), as well as using spectroscopy techniques (Belchior *et al* 2016, Dias *et al* 2018, Belchior *et al* 2019).

The VIP scores obtained in binary model and multiclass model are shown, respectively, in figures 4.4S and 4.6S in the supplementary material. In general, the VIP scores were very distinctive among the roasting degrees and sample shape, beans or ground beans, and can be attributed mainly to roasting related compounds originated from Maillard and caramelization reactions, including the formation of melanoidins, carbohydrate degradation, and reactions with proteins (Barbin *et al* 2014, Santos *et al* 2016). Moreover, other coffee compounds, such as chlorogenic acids, lipids, water, alcohols, caffeine, sugars, and phenolic acids, have also contributed to modeling and may be related to final concentrations after roasting (Ribeiro *et al* 2011, Barbin *et al* 2014, Monteiro *et al* 2018). In the dark roasting process, the thermo stable precursors present in the green coffee are more frequently degraded than in the light and medium roasting degrees (Farah *et al* 2005, 2006).

4.4 Conclusion

NIR spectroscopy in combination with multivariate classification based on PLS-DA is a reliable, alternative, and versatile method to classify coffee beans and ground coffees based on their roasting degree, mainly when they exhibit light or dark roasting degrees. This study has shown that the medium roasting degree caused ambiguity in the final coffee characteristics and affects its classification rate, which shows low sensitivity and specificity. The results of quality classification show that sensitivity and specificity vary insignificantly with the PLS-DA versions. The results for multiclass classification are very close to the binary version, and are easier and advantageous for practical implementation in the industry. Moreover, multiclass PLS-DA is no more complex than binary model classification and reveals the actual multiclass discrimination of the data.

Acknowledgments

We thank the Federal University of Lavras (UFLA), Coffee Technology and Quality Center (INOVACAFÉ), Food Science Department (UFLA), and the funding agencies CNPq, FAPEMIG, and CAPES for the financial support. We thank the producers Maria Helena Brunelli and Pedro Brunelli for providing the coffee samples. Patrícia Valderrama thanks Conselho Nacional de Desenvolvimento Científico e Tecnológico (CNPq, process 306 606/2020–8).

Credit authorship contribution statement

Fabiana de Carvalho Pires: investigation, formal analysis, data curation, and writing the original draft. Rosemary Gualberto Fonseca Alvarenga Pereira: funding acquisition, resources conceptualization, supervision, and review and editing. Michel Rocha Baqueta: data curation, formal analysis, investigation, methodology, supervision, and writing the original draft. Patrícia Valderrama: software, supervision, and review and editing. Roney Alves da Rocha: data curation, conceptualization, supervision, and review and editing.

Declaration of competing interest

The authors declare that they have no known competing financial interests or personal relationships that could have appeared to influence the work reported in this paper.

References

Agtron Inc 2010 *Coffee Roast Analyzer* (http://agtron.net/pdf/Mbas2om.pdf)

Agtron S C A A 2010 *Roast Color Classification System*

Alessandrini L, Romani S, Pinnavaia G and Rosa M D 2008 Near-infrared spectroscopy: an analytical tool to predict coffee roasting degree *Anal. Chim. Acta* **625** 95–102

Alves F C G B S and Valderrama P 2015 Ultraviolet spectroscopy and supervised pattern recognition methods for authentication of transgenic and non-transgenic soybean oils *Anal. Methods* **7** 9702–6

Baqueta M R, Coqueiro A, Março P H and Valderrama P 2021 Multivariate classification for the direct determination of cup profile in coffee blends via handheld near-infrared spectroscopy *Talanta* **222** 121526

Barbin D F, Felicio A L, de S M, Sun D W, Nixdorf S L and Hirooka E Y 2014 Application of infrared spectral techniques on quality and compositional attributes of coffee: an overview *Food Res. Int.* **61** 23–32

Barker M and Rayens W 2003 Partial least squares for discrimination *J. Chemom.* **17** 166–73

Belchior V, Botelho B G, Oliveira L S and Franca A S 2019 Attenuated Total Reflectance Fourier Transform Spectroscopy (ATR-FTIR) and chemometrics for discrimination of espresso coffees with different sensory characteristics *Food Chem.* **273** 178–85

Belchior V, Franca A S and Oliveira L S 2016 Potential of diffuse reflectance infrared Fourier transform spectroscopy and chemometrics for coffee quality evaluation *ETP Int. J. Food Eng.* **2** 1–8

Bertone E, Venturello A, Giraudo A, Pellegrino G and Geobaldo F 2016 Simultaneous determination by NIR spectroscopy of the roasting degree and Arabica/Robusta ratio in roasted and ground coffee *Food Control* **59** 683–9

Borém F M, Figueiredo L P, Ribeiro F C, Taveira J H S, Giomo G S and Salva T J G 2016 The relationship between organic acids, sucrose and the quality of specialty coffees *Afr. J. Agric. Res.* **11** 709–17

Brereton R G and Lloyd G R 2014 Partial least squares discriminant analysis: taking the magic away *J. Chemom.* **28** 213–25 https://doi.org/file:///C:/Users/Usuario/Downloads/brereton2014 (1).pdf

Catelani T A, Santos J R, Páscoa R N M J, Pezza L, Pezza H R and Lopes J A 2018 Real-time monitoring of a coffee roasting process with near infrared spectroscopy using multivariate statistical analysis: a feasibility study *Talanta* **179** 292–9

Craig A P, Botelho B G, Oliveira L S and Franca A S 2018 Mid infrared spectroscopy and chemometrics as tools for the classification of roasted coffees by cup quality *Food Chem.* **245** 1052–61

de Santana F B, Gontijo L C, Mitsutake H, Mazivila S J, de Souza L M and Borges Neto W 2016 Non-destructive fraud detection in rosehip oil by MIR spectroscopy and chemometrics *Food Chem.* **209** 228–33 de Santana F B, Souza A M, Almeida M R, Breitkreitz M C, Filgueiras P R, Sena M M and Poppi R J 2020 Experimento didático de quimiometria para classificação de óleos vegetais comestíveis por espectroscopia no infravermelho médio combinado com análise discriminante por mínimos quadrados parciais: um tutorial, parte V *Quim. Nova* 1–11

Debona D G, Catarina E, Schwengber C, Guarçoni R C, Moreira T R, Pereira L L and Moreli A P 2021 Sensory analysis and mid-infrared spectroscopy for discriminating roasted specialty coffees.

Dias R C E, Valderrama P, Março P H, dos Santos Scholz M B, Edelmann M and Yeretzian C 2018 Quantitative assessment of specific defects in roasted ground coffee via infrared-photoacoustic spectroscopy *Food Chem.* **255** 132–8

Esteban-Díez I, González-Sáiz J M and Pizarro C 2004 Prediction of sensory properties of espresso from roasted coffee samples by near-infrared spectroscopy *Anal. Chim. Acta* **525** 171–82

Farah A, De Paulis T, Trugo L C and Martin P R 2005 Effect of roasting on the formation of chlorogenic acid lactones in coffee *J. Agric. Food Chem.* **53** 1505–13

Farah A, Monteiro M C, Calado V, Franca A S and Trugo L C 2006 Correlation between cup quality and chemical attributes of Brazilian coffee *Food Chem.* **98** 373–80

Ferreira M M C 2015 *Quimiometria: conceitos, métodos e aplicações* 1st edn (Editora da Unicamp).

Fioresi D B, Pereira L L, Catarina da Silva Oliveira E, Moreira T R and Ramos A C 2021 Mid infrared spectroscopy for comparative analysis of fermented arabica and robusta coffee *Food Control* **121**

Geladi P, MacDougall D and Martens H 1985 Linearization and scatter-correction for near-infrared reflectance spectra of meat *Appl. Spectrosc.* **39** 491–500

Giraudo A, Grassi S, Savorani F, Gavoci G, Casiraghi E and Geobaldo F 2019 Determination of the geographical origin of green coffee beans using NIR spectroscopy and multivariate data analysis *Food Control* **99** 137–45

Golay A S M J E 1964 *Smoothing and differentiation of data by simplified least squares procedures***36** 1627–39

Hernández J A, Heyd B and Trystram G 2008a On-line assessment of brightness and surface kinetics during coffee roasting *J. Food Eng.* **87** 314–22

Hernández J A, Heyd B and Trystram G 2008b Prediction of brightness and surface area kinetics during coffee roasting *J. Food Eng.* **89** 156–63

Kennard A R W and Stone L A 1969 Computer aided design of experiments**11** 137–48

López M I, Callao M P and Ruisánchez I 2015 A tutorial on the validation of qualitative methods: from the univariate to the multivariate approach *Anal. Chim. Acta* **891** 62–72

Marquetti I, Link J, Lemes A L G, Scholz M B, dos S, Valderrama P and Bona E 2016 Partial least square with discriminant analysis and near infrared spectroscopy for evaluation of geographic and genotypic origin of arabica coffee *Comput. Electron. Agric.* **121** 313–9

Monteiro P I *et al* 2018 Comparison between proton transfer reaction mass spectrometry and near infrared spectroscopy for the authentication of Brazilian coffee : a preliminary chemometric study *Food Control* **91** 276–83

Nansen C, Singh K, Mian A, Allison B J and Simmons C W 2016 Using hyperspectral imaging to characterize consistency of coffee brands and their respective roasting classes *J. Food Eng.* **190** 34–9

Pereira L L, Guarçoni R C, Pinheiro P F, Osório V M, Pinheiro C A, Moreira T R and ten Caten C S 2020 New propositions about coffee wet processing: chemical and sensory perspectives *Food Chem.* **310** 125943

Pires F, de C, Pereira R G F A, Baqueta M R, Valderrama P and Alves da Rocha R 2021 Near-infrared spectroscopy and multivariate calibration as an alternative to the Agtron to predict roasting degrees in coffee beans and ground coffees *Food Chem.* **365** 130471

Poisson L, Blank I, Dunkel A and Hofmann T 2017 The chemistry of roasting-decoding flavor formation In *The Craft and Science of Coffee* (Amsterdam: Elsevier)

Pomerantsev A L and Rodionova O Y 2018 Multiclass partial least squares discriminant analysis: taking the right way—a critical tutorial *J. Chemom.* **32** 1–16

Ribeiro J S, Ferreira M M C and Salva T J G 2011 Chemometric models for the quantitative descriptive sensory analysis of Arabica coffee beverages using near infrared spectroscopy *Talanta* **83** 1352–8

Santos J R, Viegas O, Páscoa R N M J, Ferreira I M P L V O, Rangel A O S S and Lopes J A 2016 In-line monitoring of the coffee roasting process with near infrared spectroscopy: measurement of sucrose and colour *Food Chem.* **208** 103–10

Santos K M, Moura M F V, Azevedo F G, Lima K M G, Raimundo I M and Pasquini C 2012 Classification of Brazilian coffee using near-infrared spectroscopy and multivariate calibration *Anal. Lett.* **45** 774–81

Specialty Coffee Association (SCA) 2003 *Protocols and Best Practices* (https://sca.coffee/research/protocols-best-practices)

Valderrama L and Valderrama P 2016 Nondestructive identification of blue pen inks for documentoscopy purpose using iPhone and digital image analysis including an approach for interval confidence estimation in PLS-DA models validation *Chemometr. Intell. Lab. Syst.* **156** 188–95

Yergenson N and Aston D E 2020 Online determination of coffee roast degree toward controlling acidity *J. Near Infrared Spectrosc.* **28** 175–85

IOP Publishing

Advances in Spectroscopic Analysis of Food and Drink

Ashutosh Kumar Shukla

Chapter 5

The employment of infrared spectroscopy for the quality assurance of coffee

Lestyo Wulandari, Gunawan Indrayanto, Mohammad Yuwono and Abdul Rohman

Coffee is the world's most popular beverage after water, has become an essential commodity worldwide, and is among the most important raw materials in international trade. The large amount of coffee that is consumed has led to the development of quality control to meet the customer's requirements. Numerous analytical methods based on physical, chemical, and biological properties have been proposed, developed, and validated for the quality assurance of coffee, including spectroscopic- and chromatographic-based methods. Since it is a fingerprinting analytical technique, is non-destructive, and is simple in operation, infrared (IR) spectroscopy, either in near-infrared (NIR) or mid-infrared (MIR) regions, has emerged as a powerful analytical method for the quality assurance of coffee. The response obtained from IR spectral measurement involves large data sets, and therefore the use of multivariate data analysis or chemometrics is unavoidable. The combination of IR spectroscopy and chemometrics of pattern recognition has been successfully applied for discrimination and classification of coffee with different grades. In addition, IR spectroscopy coupled with multivariate calibrations offers an accurate and precise method for the prediction of chemical components in coffee, which is intended for the quality control of coffee products.

5.1 Introduction

Coffee is a plantation crop that has high economic value. Robusta coffee dominates production in Indonesia—roughly 60% of coffee production in Indonesia is Robusta coffee, while the remaining 40% is Arabica coffee [1]. Indonesia's 2022–23 coffee production is forecast to increase by 7% from the previous year to 11.35 million (60 kg) bags, based on favorable weather in southern Sumatra. The improved demand following the easing of COVID19 pandemic-related restrictions is expected to increase domestic coffee consumption to 4.8 million bags in 2022–23 [2]. Coffee is

doi:10.1088/978-0-7503-5573-5ch5 © IOP Publishing Ltd 2023

an agricultural product that can be processed into drinks that have a delicious aroma and taste, it also has potential to be used as a stimulating drug and as a drowsiness inhibitor [3]. Consumers drink coffee for its energetic and therapeutic effects. Coffee consumption is still price-driven, but many consumers are interested in purchasing coffee with associated health claims (a relevant minority of consumers believe that coffee could have positive health effects). Brand identification is especially important in the coffeehouse market, drinking a specific coffee brand, e.g., Starbucks, represents a status symbol and way of life for consumers [4].

According to the International Trade Centre's Coffee Exporter's Guide [5], coffee is usually graded on the basis of one of more of the following criteria: (1) altitude and region; (2) botanical type; (3) preparation (wet or dry process, washed or natural); (4) bean size, shape, or color; (5) number of defects; (6) roast appearance and cup quality (flavor, characteristics, cleanliness); and (7) bean density. According to the Indonesian National Standard (SNI) 01-2907-2008 [6], the quality standard requirements of coffee beans consisted of some criteria, namely the condition/description, clasification (robusta, arabica), testing of water content,bean size, no live insects, impurity, and bean defects.,. Coffee quality is also determined by taste testing through sensory analysis via panelists, sensory assessment parameters of powder aroma, beverage aroma, acidity, bitterness, taste, astringency, residual taste, and overall quality [7]. Quality assessment of coffee should be carried out in six different aspects of the production, i.e., pre-harvest, post-harvest, coffee bean storage, coffee bean processing, coffee powder products, and coffee beverage products. The complex relationship between the quality of coffee seeds and their volatile components in the six different aspects has also been discussed [8]. Regardless of its nature, forms of growth, industrialization, and preparation, the final quality of beverages and their sensorial evaluation depend directly on the chemical composition of its roasted beans, e.g., coffee contains some secondary metabolites such as caffeine, trigonelline, and chlorogenic acids [9]. The sensory (aroma, flavor, roast) aspects of coffee products can be affected by their primary and secondary metabolites contents.

Currently, USP45-NF40 describes another type of coffee product, i.e. coffee fruit dry extract. The extract is prepared from the dried semi-ripened light red entire fruit of *Coffea arabica* L. (Fam. Rubiaceae) by extraction with hydroalcoholic mixtures. It contains not less than (NLT) 40% of caffeoylquinic acids, calculated as the sum of neochlorogenic acid, chlorogenic acid, cryptochlorogenic acid, isochlorogenic acid B, isochlorogenic acid A, and isochlorogenic acid C on the dried basis; and NLT 15% of chlorogenic acid on the dried basis [10]. According to the website of Dave Asprey, coffee extracts can be used as a natural nootropic (brain-boosting) supplement due to their capability to boost brain-derived neurotrophic factors. The extracts are available as food supplements [11].

5.2 Quality assessment of coffee

The standard specifications for roasted coffee beans and roasted ground coffee of the SNI of powdered coffee [12] and Uganda [13] are listed in table 5.1. The standard

Table 5.1. Standard quality requirement of powdered coffee of Indonesia and Uganda[1].

Test parameter	SNI 01-3542-2004,	Uganda standard
Organoleptic (color, smell, texture)	Normal	NA[2]
Moisture, % by weight	7	5.0
Total ash, % by weight	5	2.0–5.0
Acid insoluble ash, % by weight	NA[2]	1.0
Water-soluble matter, % by weight	20–36	25–32
Alkalinity of water-soluble ash in milliliters of 0.1 N hydrochloric acid per gram of material	5.7–6.4	3.5–7.0
Caffeine, % by weight	0.9–2.0	0.8 0.1 for decaffeinated
Total plate count, cfu g^{-1}	10^6	10^3
Coliforms cfu g^{-1}	NA[2]	<10
Salmonella spp per 25 g	NA[2]	Absent
Yeast and *molds* cfu g^{-1}	10^4	10^2
Lead (mg kg^{-1})	2.0	0.5
Cadmium (mg kg^{-1})	NA[2]	0.1
Cu (mg kg^{-1})	30.0	NA[2]
Zn (mg kg^{-1})	40.0	NA[2]
Sn (max mg kg^{-1})	40.0	NA[2]
Hg (mg kg^{-1})	0.03	NA[2]
As (mg kg^{-1})	1.0	NA[2]

[1] Maximum permitted concentrations:[2] Not available

requirement for the coffee bean (*Coffea robusta* and *C. arabica*) is described by SNI 01-2907-2008 [6], while ASEAN Stan 31:2013 defines the standards of green coffee bean of the commercial varieties grown from *C. arabica*, *C. robusta* (*Coffea canephora*), *Coffea liberica*, and *Coffea excelsa* [14]. The standards for coffee beans generally consist of guidelines and specifications for the sampling method, general quality, size tolerance, moisture content, uniformity, maximum defect, contaminants, and packaging.

The International Coffee Organization, which has 23 member countries [5], described general quality standard and specification for green coffee as follows: altitude, type (Robusta/Arabica), process preparation, bean (size, shape, color, density), moisture, odor, number of defects, cup quality testing, sensorial (aroma, roast, flavor), and number of grades. Each country has its own quality standards: 12 countries have described sensory evaluations, but these are not described by 11 other countries. Unfortunately, only a few countries have described the standard quality requirements of the secondary metabolites content, i.e., caffeine and chlorogenic acids. Some toxic contaminants have been detected in coffee, such as ochratoxin A, acrylamide, aflatoxin, polycyclic aromatic hydrocarbons, yeast, and pesticides [15]. A review article on the possible contamination of pesticides in coffee, including their methods of analysis, has recently been published [16]. The general chapter of the current USP-NF <561> describes the specification limit of more than 60 pesticides in botanical products [17]. The determination of metals-microbiological-contaminations is also crucial for the consumer's safety [12].

The International Organization for Standardization (ISO) has published quality testing standards and guidelines for coffee i.e. ISO/TC34/SC15, including sampling, storage, sensory analysis, determination of caffeine, psychochemical testing etc [18]. According to Mermelstein [19], the Association of Official Agricultural Chemists (AOAC) has published 29 official methods for quality testing of coffee, e.g., macroscopic examinations, physicochemical properties (moisture, ash, acidity, fiber etc), metabolites determination (caffeine, chlorogenic acids, sugars), and contaminants (mycotoxin, ochratoxin A). The URL of the Institute of Food Technologists (IFT) has listed 1188 articles of coffee quality that can be downloaded from their link [20]. The Indonesian Coffee and Cocoa Research Institute, Jember, Indonesia described four score criteria for the sensory evaluation of cup coffee, i.e., (1) outstanding (score 9.00–9.75), (2) excellent (score 8.00–8.75), (3) very good (score 7.00–7.75), and (4) good (score 6.00–6.75) (personal communication with the technical manager of the Institute, March 2023). The methods of the sensory evaluations used by this institute are based on the methods published by Specialty Coffee Association (SCA), i.e., Standard 320-2021 Home Coffee Grinders: Specifications and Test Methods [21]. It is already known that sensory plant products, e.g., teas, coffee, fruits, juices etc, will be affected by their whole metabolite contents (primary and secondary). Therefore, for quality assessment of the coffee, a determination of all main metabolites must be performed, instead of assessing caffeine and chlorogenic acids only.

Various chemical methods, i.e., ultraviolet (UV) spectroscopy, thin layer chromatography (TLC), high performance liquid chromatography ultraviolet (HPLC-UV), high performance liquid chromatography diode array detector (HPLC-DAD), liquid chromatography mass spectrometry (LC-MS), liquid chromatography with tandem mass spectrometry (LC-MS/MS), GC-NPD, electrochemical methods, and IR spectroscopy, have been used to determine the caffeine content of coffee [22]. Application of chromatographic methods for the analysis of various components of coffee has been reviewed by Yashin *et al* [23]. TLC has been used for the determination of chlorogenic acid, polyphenols, and antioxidants in green coffee [24]. The adulteration of coffee has been determined using HPLC combined with chemometrics analysis [25]. A rapid and simple determination of caffeine in teas, coffees, and eight beverages using GC-NPD was reported by Sereshti and Samadi [26]. Structural characterization of native and oxidized procyanidins (condensed tannins) from coffee pulp (*C. arabica*) using phloroglucinolysis and thioglycolysis has been determined using HPLC-ESI-MS [27]. Determination of caffeine and nicotinic acid in coffee has been performed using a HPLC-UV/Vis detector [28]. Determination of caffeine in coffee products [29] and methylcafestol in roasted coffee products [30] have also been reported. HPLC electrochemical determination of caffeine content in Ethiopian coffee samples using lignin modified glassy carbon electrode has been reported [31]. Chemometric approaches to discriminate coffees based on nuclear magnetic resonance spectroscopy, chromatography, IR or Raman spectroscopy, and array sensors/indicators has been reviewed and discussed [32]. The drawbacks of chromatography and electrochemical methods are time-consuming sample preparation and/or extraction prior to chromatographic measurements, or preparation of a specific electrode. Therefore, the application of direct spectroscopic and non-destructive methods such as NIR spectroscopy, Fourier transform infrared (FT-IR) spectroscopy, mass spectrometry (MS), and nuclear magnetic resonance should be further considered [33]. Due to their relatively low operation costs, the application of NIR and FT-IR is preferred.

5.3 Infrared spectroscopy and its applications for food analysis

Infrared spectroscopy includes NIR spectroscopy and MIR spectroscopy. According to the general chapters of the current United States Pharmacopoeia - National Formulary (USP-NF), NIR spectroscopy covers the electromagnetic field wavelength range from 780 to 2500 nm, with the frequencies expressed in wavenumbers (about 12,821–4000 cm^{-1}), while MIR spectroscopy is an instrumental method used to measure the absorption of electromagnetic radiation over the wavenumber range between 4000 and 400 cm^{-1} (corresponding to wavelengths between 2.5 and 25 μm) [34, 35]. These two general chapters of the current USP described also the methods of qualifications (installation qualification, operational qualification, performance qualification, and wavelength accuracy test) of the IR spectrophotometers. Because most of the MIR spectra are measured with FT-IR spectrophotometers [35], the acronym FT-IR will be used to describe MIR spectroscopy in this review. The theory of FT-IR spectroscopy, infrared processes, and sampling techniques, e.g., transmission, attenuated total reflection (ATR), diffuse reflectance etc, has been described and discussed in our previous chapter [36]. The general chapter of the current USP45-NF40 <856> [37] has described the theory and practice of the NIR spectroscopy in detail. NIR and FT-IR spectroscopy can be applied for various quality assessments of drugs and food related products, i.e., qualitative identification of active compounds, herbs, food materials, excipients, packaging materials, quantitative analysis of compounds (active ingredients, water, hydroxyl values etc), physical characterizations (polymorphisms, particle sizes, hardness etc), and in-line processing monitoring [38]. However, the analyte signal of the IR spectroscopy often does not display the appropriate degree of linearity or signal-to-noise as in chromatographic methods. Moreover, no single variable is directly and uniquely related to the target concentration. Therefore, the application of chemometrics, such as principle component analysis (PCA), partial least square (PLS), PLS-DA, soft independent modeling of class analogy, and others need to be combined with IR spectroscopic analysis methods [38]. A review article on the overview of variable selection methods in multivariate analysis using NIR spectroscopy has been published by Yun *et al* [39].

We have previously described and discussed the use of FT-IR spectroscopy combined with chemometrics analysis in general food composition analysis [36]. A summary of the application of the combination of chemometrics and FT-IR spectroscopy for the analysis of foods (coffee, diaries, honey, olive oils, and wine) has also been published [40]. The application of the combination of FT-IR and multivariate analysis (MVA) for the quality control of coffee, cocoa, and commercially important spices has recently been reviewed [41]. The application of FT-IR spectroscopy and chromatography in combination with chemometrics for the quality control of olive oil has been reviewed and discussed [42]. Quality assessment of fats and oils using FT-IR spectroscopy combined with chemometrics analysis has also been published recently [43]. Feng *et al* [44] discussed the applications of UV spectroscopy, IR spectroscopy, and hyperspectral imaging with machine learning techniques to identify food varieties and geographical origins.

5.4 Application of the combination of MVA and infrared spectroscopy for the quality assessment of coffee

Some recent publications on the application of NIR and FT-IR spectroscopy for the quality assessments of coffee are summarized in table 5.2.

As shown in table 5.2, direct infrared spectroscopy can be applied for the quality assessment of coffee, including determination of species, adulteration, physicochemical properties, roasting processes, sensory evaluation, and metabolites content. Most of the cited publications used chemometric methods to evaluate the data, only a few of them used univariate methods. The advantages of the direct spectroscopic methods are that they nn-destructive, they do not require complex sample preparations, they are relatively fast, and they are low cost. Unfortunately, these methods cannot be applied to predict or determine the identity of the metabolites. In this case, more sophisticated instruments, such as LC/GC-MS or MS/MS, are needed. The reliability of the analysis using IR spectroscopy depends on the availability of authentic and stable standards (chemical compounds, raw materials, and extracts), and whether the methods have been fully validated according to newest official guidance.

It can be seen in table 5.2 that most of the cited publications did not completely report their validation methods as requested by the formal guidance, e.g., general chapter <1039> Chemometrics of the current USP-NF [67]. For quantitative chemometric analysis, validation parameters such as accuracy, repeatability, intermediate precisions, and ranges should be evaluated according to official guidance, e.g., chapter <1225> of the current USP-NF [68] or chapter <1381> of the e-Indonesian Pharmacopoeia VI [69]. Validation of quantitative chemometrics methods should performed in two steps: (1) validation of the chemometrics model using RMSEC, RMSEP etc, and (2) validation of the method using accuracy, precision etc. The validation of the chemometrics methods has been described and discussed by Noviana *et al* [33]. The interested reader can also refer to the AOAC appendix K Guidelines for the general validation methods of dietary supplements and/or botanical related products [70].

Table 5.2. Application of IR spectroscopy for the quality assessment of coffee.

Sample (objective of the work)	Instrument	Sample measurements	MVA methods or other statistic evaluation	Validation methods of evaluation	Results	Reference
Roasting coffee (monitoring coffee roasting process).	FT-NIR using InGaAs detector and acquired through a stainless-steel diffuse reflectance probe	The probe was placed outside the coffee roaster at approximately 2 mm (64 scans), resolution 8 cm (10,000–4000 cm^{-1}).	Multivariate statistical process control methodology based on PCA models.	Validation and testing sets were randomly performed; specificity and sensitivity were determined.	NIR using MVA can be used to monitor the coffee roasting process in real-time.	[45]
Roasting Arabica coffee (*in situ* monitoring coffee roasting cracks).	FT-NIR with *in situ* diffuse reflectance detector (550 µm with 0.5 m long low hydroxyl ions fiber optic cable)	Two 50 W halogen bulbs placed at a 45° from the fiber probe, NIR spectrum was captured every 10 s during roasting.	PLS regression.	RMSEC, RMSEP, R^2, and range error ratio were determined.	NIR models can be used to predict the start and end time of the first and second crack events.	[46]
Green coffee, Arabica and Robusta coffee (sensory evaluations in coffee cupping).	NIR with a micro-diffuse reflectance detector	NIR spectra were obtained from the ground coffee (900–1650 nm).	PLS models.	RMSEC, RMSEP, residual prediction deviation (calibration/ validation), accuracy, linearity, sensitivity, LOD, and LOQ were determined.	NIR combined chemometrics could be used to predict the sensorial attributes of roasted and ground samples.	[47]

(*Continued*)

Table 5.2. (*Continued*)

Sample (objective of the work)	Instrument	Sample measurements	MVA methods or other statistic evaluation	Validation methods of evaluation	Results	Reference
Arabica coffee from 14 countries (prediction of specialty coffee flavor).	NIR (no detailed information)	Ground sample was scanned 32 times (700–2500 nm at 2 nm intervals).	PCA, ML, and DCNN.	Evaluated by cross-validation; accuracy[1] and recall[1] were determined.	NIR (ground coffee) combined with ML and DCNN can be used to predict specialty coffee flavors.	[48]
Arabica green coffee and roasting beans (predicting roasting degrees of coffee beans and ground coffees).	NIR equipped with 45° ZnSe crystal.	Spectra were recorded from ground samples (4000–10,000 cm^{-1}, resolution 4 cm^{-1}, 64 scans/samples).	The variable importance projections of the PLS models were determined.	Evaluated by cross-validation; RMSEC, RMSEP, R^2, accuracy, LOD, and LOQ were determined.	NIR combined with MVA can be applied to predict roasting profiles in roasted whole coffee beans and ground coffees.	[49]
Arabica coffee bean samples (tracing the geographical origins of coffee bean).	NIR (no detailed information)	Spectra were measured at 400–2498 nm, 2-nm intervals, 32 scans; psychochemical properties were also determined.	PCA, SOM, and artificial neural network.	SOM discriminant index[1] is reported.	NIR coupled with SOM can be applied to identify the origin of coffee collected from different geographical origins.	[50]
Green coffee beans samples (comparison UV and NIR determination of caffeine and chlorogenic acid).	FT-NIR, rotating sample cup, and UV spectrophoto meter	Spectra (diffuse reflectance) measured at 1000–2500, 2, 64 scans; UV: $\lambda = 276$ nm (caffeine), $\lambda = 324$ nm (chlorogenic acid).	PCA, PLS-DA, LDA, and latent variables were determined.	R^2, RMSEC, RMSEP, and accuracy were determined.	Both methods can be applied with acceptable accuracy.	[51]

Unroasted green coffee beans, and roasted coffee (sensory and quantification total phenolics).	NIR (no detailed information)	Spectra were obtained from whole grains (4000–12,000 cm^{-1}, resolution 8 cm^{-1}, triplicate).	PCA and PLS regression.	Evaluated by cross-validation, RMSEC, RMSEP, and R^2 were determined.	Quality can be estimated from green beans, which makes the process more practical, quick, and economical.	[52]
Coffee beans with known differences in cup quality and roasting degree (direct determination of cup profile in coffee blends).	Handheld portable microNIR™ 1700	Roasted ground coffees were directly analyzed on microNIR (reflectance; 906–1676 nm, 32 scans).	PLS-DA and PLS regression.	Receiver operating characteristic, RMSECV, RMSEC, RMSEP, Sensitivity[1], and Specificity[1] were determined.	The sensitivity and specificity of the model obtained ranged from 91%–100%, 84%–100%, and 73%–95% (for training, prediction, and internal cross-validation sets, respectively).	[53]
Green coffee beans, roasted coffee beans (correlating the composition of green beans with the sensory quality).	NIR equipped with diffuse reflectance detector	NIR spectra were collected at 1100–2500 nm, interval 2 nm^{-1}.	HCA and PCA.	Pearson correlation coefficient ($p<0.05$).	The profile of the composition of the crude coffee can be indicative of the sensory quality of the coffee brews.	[54]
Various preparations of Arabica and *Robusta coffee* (comparative analysis of fermented coffee).	FT-IR equipped with diamond ATR and ZnSe reflectance detector	FT-IR spectra were collected at 4000–630 cm^{-1}, resolution 1 cm^{-1}, 256 scans	Student's t-test and Mann Whitney test[1].	Significance test ($p < 0.05$).	Fermentation was demonstrated to be the most recommendable processing technique for the Robusta variety.	[55]

(Continued)

Table 5.2. (*Continued*)

Sample (objective of the work)	Instrument	Sample measurements	MVA methods or other statistic evaluation	Validation methods of evaluation	Results	Reference
Arabica green coffee (determination of coffee adulteration by corn).	Differential scanning calorimeter[1] and FT-IR with reflection monolithic diamond device	Each spectrum had an average of 64 scans from 525–4000, 0.4 cm^{-1} resolution, 1.	PCA and PLS	RMSEC, RMSECV, RMSEP, and R^2 (cross-validation, calibration, and prediction) were determined.	PCA is able to discriminate adulterated from unadulterated coffee samples at levels below 1%. PLS can be used to estimate the concentrations of adulterants.	[56]
Coffee brands present in the Spanish market (determination of bioactive compounds in coffee beans).	NIR equipped with hyperspectral device.	Samples were extracted from plastic bags and placed on a polyethylene plastic tray. They were then placed under a hyperspectral device for image acquisition.	ANOVA, PCA, PLS regression, and modified partial least square regression.	RMSEC, RMSECV, RMSEP, R^2, RPD, and bias were determined.	NIR can be applied for quality control of coffee beans to determine bioactive compounds.	[57]
Arabica green coffee (determination of pH and acidity in green coffee).	FT-NIR in reflectance mode	Sample was analyzed in a glass bottle of 125 ml in constant rotation at 800–2600 nm, resolution 16 cm^{-1}, 64 scans.	PLS regression.	RMSEC, RMSECV, RMSEP, R^2, and RPD were determined.	NIR can be applied to determine physicochemical properties (pH and acidity).	[58]

Arabica, Canephora green coffee, and commercial coffee (classification of roasted coffees by cup quality).	ATR-FT-IR equipped with ZnSe cell and DLATGS detector	Ground and roasted coffee was placed in the sampling accessory, pressed, and then measured (4000–600 cm^{-1}, resolution 4 cm^{-1}, 20 scans).	PCA and PLS-DA.	Sensitivity[1] and selectivity[1] were determined.	FT-IR combined with PLS-DA can be used for the quality evaluation of arabica coffees regardless of roasting conditions.	[59]
Arabica coffees were collected at different altitudes and then roasted using different temperatures (discriminating roasted specialty coffees).	ATR-FT-IR with diamond accessory (45° reflection angle, 0.001 m in diameter, active area of 200 μm)	Tablets (without Kbr) were made from ground coffee and then placed on the ATR crystals.	HCA and PCA.	NA	FT-IR was able to identify the classes of compounds responsible for the separation of the roasting profiles for coffees grown at different altitudes.	[60]
Grounded coffee beans of *C. arabica* and *C. canephora* and extracted matrix coffee (predicting compounds concentrations).	NIR (no detailed information) and HPLC (for compounds determi nations)	Sample was profiled in triplicate with 256 scans in the 1100–2500 nm range and resolution of 2 nm; HPLC see Ref[1]	PLS regression and latent variable was determined.	R^2, RMSECV, and RMSEP were determined.	PLS models can be used to determine caffeine, trigonelline and 5-caffeoylquinic acid.	[61]
Roasting coffee (differentiating between the dry, semi-dry, and wet post-harvest methods using FT-IR).	ATR-FT-IR equipped with DLATGS detector	Spectra were taken in the range in the range of 4000–650 cm^{-1}, resolution 4 cm^{-1} and 16 scans	ANOVA, PCA, and LDA.	DF1[1] and DF2[1] were determined.	PCAs can be applied to discriminate the post-harvest treatments and LDA can be used in the recognition of post harvesting samples.	[62]

Table 5.2. (*Continued*)

Sample (objective of the work)	Instrument	Sample measurements	MVA methods or other statistic evaluation	Validation methods of evaluation	Results	Reference
Arabica green coffee (selective determination of caffeine and trigonelline in aqueous extract).	ATR-FT-IR with sample holder of ZnSe crystal	Samples were scanned at 1430 to 1170 cm^{-1}, quantitative was performed at 1243 cm^{-1} (caffeine) and 1212 cm^{-1} (trigonelline).	Univariate method.	Linearity, LOD, LOQ, recovery, and precision were determined; selectivity was not reported.	ATR-FT-IR methods can be used for direct determination of caffeine and trigonelline.	[63]
Commercially roasted and ground coffee (determination coffee quality using pattern recognition approaches).	FT-IR equipped with DLATGS detector and ATR sampling accessory (diamond/ ZnSe cell)	Spectra were taken in the range in the range of 4000–500 cm^{-1}, resolution 4 cm^{-1} and 16 scans.	PCA and LDA.	Sensitivity, specificity and accuracy were determined.	FT-IR combined with PCA and LDA can be applied for authentication purposes.	[64]
Commercial green coffee beans (determination of lipid and protein content).	NIR equipped with InGaAs detector	Spectra were recorded in the range 12000-4000 cm^{-1}, resolution 8 cm^{-1}, 64 scans.	PCA and PLS regression.	RMSEC, RMSECV, RMSEP, R^2, RPD, LOD, and LOQ were determined.	NIR techniques combined with PLS regression can be applied for the determination of protein and lipid content.	[65]

Green coffee of *C. arabica* and *C. canephora* (differentiate coffee based on their roasting levels using DFA).	FT-NIR (no detailed information)	Spectra were collected in diffuse reflection mode (12,500–3800 cm^{-1}), (resolution 16 cm^{-1}), scanning speed 10 kHz.	PCA, HCA, multiple correspondence analysis, and DFA.	Yield stability index[1] is calculated.	DFA can be applied to differentiate the green samples with the roasted samples. DFA can be used to determine the roasting levels of the samples.	[66]

[1] See original reference for a detailed description (seventh column of the table).

[2] DFA = detrended fluctuation analysis, InGaAs = indium gallium-arsenide, ZnSe = Zink Selenide, DLATGS = deuterated triglycine sulfate doped with L-alanine, ML = machine learning, DCNN = deep convolutional neural network, SOM = self-organizing map, LDA = linear discriminant analysis, HCA = Hierarchical cluster analysis, ANOVA = analysis of variance, RMSEC = root mean square calibration, RMSEP = root mean square error of prediction, R^2 = coefficient of determination of explained variance, LOD = limit of detection, LOQ = limit of quantification, RMSECV = root mean square error of cross validation, RPD = ratio of performance, DF = discriminant function.

5.5 Conclusions and recommendations

The ISO, SCA, IFT, AOAC, and many countries have published guidelines and standard quality requirements to ensure the quality of coffee-related products. Generally, the guidelines and the standards requirements describe the methods of sampling, sensory evaluations, psychochemical testing, assay of secondary metabolites, and the detection of toxic chemical contaminants (metabolites, pesticides, heavy metals, and microbiological). The guidelines and standards requirements that are used in different countries can differ. For the safety of coffee consumers, controlling and detecting toxic contaminants are very important. Consequently, it is recommended that the national food control authorities should sample marketed coffee for routine analysis, which can be used to detect toxic contaminants as early as possible (if any).

Chromatographic and spectroscopic methods of analysis may be applied for the quality assessment of coffee and its products. However, due to the weakness of the chromatographic method of analysis, the application of the combination of direct spectroscopic methods and chemometrics analysis should be considered. This combination of methods is relatively fast and low cost but is still accurate enough if authentic standards are available. To save time in controlling marketed coffee, it would be useful to develop, standardize, and validate chemical methods (direct spectroscopy) that can replace the sensory and psychochemical analysis of the quality of coffee.

Acknowledgement

We acknowledge the R & D Department of PT Konimex, Solo, Indonesia for permission to access the USP-NF online.

References

[1] USDA, United States Department of Agriculture Coffee Annual, Indonesia 2021, 2023 (https://www.fas.usda.gov/data/indonesia-coffee-annual-5) (accessed 25 March 2023).

[2] USDA, United States Department of Agriculture Coffee Annual, Indonesia, 2022, 2023 (https://www.fas.usda.gov/data/indonesia-coffee-annual-6) (accessed 25 March 2023).

[3] van Dam R M, Hu F B and Willett W C 2020 Coffee, caffeine, and health *N. Engl. J. Med.* **383** 369–78

[4] Samoggia A and Riedel B 2019 Consumers' perceptions of coffee health benefits and motives for coffee consumption and purchasing *Nutrients* **11**

[5] International Coffee Organization, National Quality Standards, ICC 122-12 2018 (https://www.ico.org/documents/cy2017-18/icc-122-12e-national-quality-standards.pdf) (accessed 13 March 2023).

[6] 2008 ICS 67.140.20, BSN, Badan Standardisasi Nasional, Standard Nasional Indonesia, SNI 01-2907–2008, Biji Kopi, Badan Standardisasi Nasional Indonesia

[7] Lingle T R 2011 *The Coffee Cupper's Handbook Systematic Guide to the Sensory Evaluation of Coffee's Flavor* 4th (Long Beach, CA: Specialty Coffee Association of America)

[8] Toledo P R A B, Pezza L, Pezza H R and Toci A T 2016 Relationship between the different aspects related to coffee quality and their volatile compounds *Compr. Rev. Food Sci. Food Saf.* **15** 705–19
[9] Abreu M B, Marcheafave G G, Bruns R E, Scarminio I S and Zeraik M L 2020 Spectroscopic and chromatographic fingerprints for discrimination of specialty and traditional coffees by integrated chemometric methods *Food Anal. Methods* **13** 2204–12
[10] USP45-NF40, Coffee fruit dry extract, (2023). doi: https://doi.org/10.31003/USPNF_M10035_05_01 (accessed 13 March 2023).
[11] Asprey D 2023 *Coffee Fruit Extract Is the Superfood Supplement Your Brain Needs* (https://daveasprey.com/coffee-fruit-extract-nootropic-supplement/) (accessed 13 March 2023).
[12] ICS 67 140.20, BSN, Badan Standardisasi Nasional, Standard Nasional Indonesia, SNI 01-3542-2004 *Kopi Bubuk* 2004
[13] Uganda National Bureau of Standards, Draft Uganda Standard, DEUS DEA 105 2019 (https://members.wto.org/crnattachments/2019/SPS/UGA/19_1688_00_e.pdf) (accessed 27 March 2023).
[14] ASEAN, ASEAN Standard for Coffee Bean: ASEAN Stan 31: 2013 2013 (https://asean.org/wp-content/uploads/images/Community/AEC/AMAF/UpdateApr2014/ASEAN%20standard%20for%20coffee%20bean.pdf) (accessed 24 March 2023).
[15] Natural Force, Mold, Toxins, and Pesticides: Coffee's Dirty Secret 2023 (https://naturalforce.com/blogs/nutrition/coffee-mold-toxins-pesticides) (accessed 23 March 2023).
[16] Merhi A, Kordahi R and Hassan H F 2022 A review on the pesticides in coffee: usage, health effects, detection, and mitigation *Front. Public Health* **10** accessed 23 March 2023
[17] USP45-NF40, General Chapters, General Test and Assay <561> Article of Botancal Origin 2022 (https://online.uspnf.com/uspnf/document/1_GUID-E8A1366F-9657-41FC-9EDC-C20F4BE473B6_6_en-US) (accessed 12 May 2022).
[18] ISO, Standards by ISO/TC 34/SC 15 *Coffee* 2023 (https://www.iso.org/committee/47950/x/catalogue/) (accessed 24 March 2023).
[19] Mermelstein N H 2012 Coffee quality testing *Food Technology Magazine* (https://www.ift.org/news-and-publications/food-technology-magazine/issues/2012/january/columns/food-safety-and-quality) (accessed 23 March 2023).
[20] Institute of Food Technologists 2023 *Coffee Quality* (https://www.ift.org/search#q=coffee%20quality%20&sort=relevancy) (accessed 23 March 2023).
[21] Specialty Coffee Association, (SCA) 2022 *Home Coffee Grinders: Specifications and Test Methods* (https://bit.ly/3PmBebo) (accessed 18 March 2023).
[22] Sharma H 2020 A detail chemistry of coffee and its analysis *Coffee—Production and Research* ed D T Castanheira (London: IntechOpen) 1–12
[23] Yashin A, Yashin Y, Xia X and Nemzer B 2017 Chromatographic methods for coffee analysis: a review *J. Food Res.* **6** 60
[24] Choma I M, Olszowy M, Studziński M and Gnat S 2019 Determination of chlorogenic acid, polyphenols and antioxidants in green coffee by thin-layer chromatography, effect-directed analysis and dot blot—comparison to HPLC and spectrophotometry methods *J. Sep. Sci.* **42** 1542–9
[25] Cheah W L and Fang M 2020 HPLC-based chemometric analysis for coffee adulteration *Foods* **9**
[26] Sereshti H and Samadi S 2014 A rapid and simple determination of caffeine in teas, coffees and eight beverages *Food Chem.* **158** 8–13

[27] Wong-Paz J E, Guyot S, Aguilar-Zárate P, Muñiz-Márquez D B, Contreras-Esquivel J C and Aguilar C N 2021 Structural characterization of native and oxidized procyanidins (condensed tannins) from coffee pulp (*Coffea arabica*) using phloroglucinolysis and thioglycolysis-HPLC-ESI-MS *Food Chem.* **340**

[28] Chirfa G, Merdassa Y and Gure A 2020 Salting-out assisted liquid–liquid extraction for analysis of caffeine and nicotinic acid in coffee by HPLC–UV/Vis detector *J. Anal. Test.* **4** 298–306

[29] Naegele E 2016 Determination of Caffeine in Coffee Products According to DIN 20481, Application Notes, Agilent Technologies (accessed 27 March 2023).

[30] Naegele E 2016 Determination of methylcafestol in roasted coffee products according to DIN 10779 Application Notes, Agilent Technologies (https://www.perlan.com.pl/uploaded/AppBundleEntityProductApplication/fileKey/107/5991-2853en.pdf) (accessed 27 March 2023).

[31] Amare M and Aklog S 2017 Electrochemical determination of caffeine content in ethiopian coffee samples using lignin modified glassy carbon electrode *J. Anal. Methods Chem.* **2017**

[32] Wang X, Lim L T and Fu Y 2020 Review of analytical methods to detect adulteration in coffee *J AOAC Int.* **103** 295–305

[33] Noviana E, Indrayanto G and Rohman A 2022 Advances in fingerprint analysis for standardization and quality control of herbal medicines *Front. Pharmacol.* **13**

[34] USP45-NF40 2023 General chapters, general test and assay, <856> near-infrared spectroscopy accessed 12 March 2023

[35] USP45-NF40 2023 General chapters, general test and assay, <854> mid-infrared spectroscopy accessed March 12, 2023

[36] Indrayanto G and Rohman A 2020 The use of FTIR spectroscopy combined with multivariate analysis in food composition analysis ed A K Shukla *Spectroscopic Techniques & Artificial Intelligence for Food and Beverage Analysis* (Singapore: Springer Nature) pp 25–51

[37] USP45-NF40 2023 General chapters, general test and assay, <1856> near-infrared spectroscopy, theory and practice accessed March 12, 2023

[38] Wulandari L, Idroes R, Noviandi T R and Indrayanto G 2022 Application of chemometrics using direct spectroscopic methods as a QC tool in pharmaceutical industry and their validation ed A A Al-Majed *Profile of Drug Substances, Excipients, and Related Methodology* (Cambridge: Academic Press, Elsevier) pp 330–79

[39] Yun Y H, Li H D, Deng B C and Cao D S 2019 An overview of variable selection methods in multivariate analysis of near-infrared spectra *TrAC—Trends Anal. Chem.* **113** 102–15

[40] Mendes E and Duarte N 2021 Mid-infrared spectroscopy as a valuable tool to tackle food analysis: a literature review on coffee, dairies, honey, olive oil and wine *Foods* **10** 1–32

[41] Castillejos-Mijangos L A, Acosta-Caudillo A, Gallardo-Velázquez T, Osorio-Revilla G and Jiménez-Martínez C 2022 Uses of FT-MIR spectroscopy and multivariate analysis in quality control of coffee, cocoa, and commercially important spices *Foods* **11**

[42] Indrayanto G and Rohman A 2022 Application of FTIR spectroscopy and chromatography in combination with chemometrics for the quality control of olive oil ed A K Shukla *Food Chemistry, Function and Analysis No. 32 Advanced Spectroscopic Techniques for Food Quality* (London: The Royal Society of Chemistry) pp 133–79

[43] Rohman A and Indrayanto G 2023 Fourier transform infrared spectroscopy combined with multivariate analysis of fats and oils ed A K Shukla *Food Quality Analysis* (Amsterdam: Elsevier Inc.) pp 49–70

[44] Feng L, Wu B, Zhu S, He Y and Zhang C 2021 Application of visible/infrared spectroscopy and hyperspectral imaging with machine learning techniques for identifying food varieties and geographical origins *Front Nutr.* **8**

[45] Catelani T A, Santos J R, Páscoa R N M J, Pezza L, Pezza H R and Lopes J A 2018 Real-time monitoring of a coffee roasting process with near infrared spectroscopy using multivariate statistical analysis: a feasibility study *Talanta* **179** 292–9

[46] Yergenson N and Aston D E 2020 Monitoring coffee roasting cracks and predicting with *in situ* near-infrared spectroscopy *J. Food. Process Eng.* **43**

[47] Baqueta M R, Coqueiro A and Valderrama P 2019 Brazilian coffee blends: a simple and fast method by near-infrared spectroscopy for the determination of the sensory attributes elicited in professional coffee cupping *J. Food Sci.* **84** 1247–55

[48] Chang Y T, Hsueh M C, Hung S P, Lu J M, Peng J H and Chen S F 2021 Prediction of specialty coffee flavors based on near-infrared spectra using machine—and deep-learning methods *J. Sci. Food Agric.* **101** 4705–14

[49] de F, Pires C, Pereira R G F A, Baqueta M R, Valderrama P and Alves da Rocha R 2021 Near-infrared spectroscopy and multivariate calibration as an alternative to the Agtron to predict roasting degrees in coffee beans and ground coffees *Food Chem.* **365**

[50] Wongsaipun S, Theanjumpol P, Funsueb S, Kittiwachana S, Mai Sci C J, Wongsaipun S, Muenmanee N and Boonyakiat D 2021 Application of artificial neural network for tracing the geographical origins of coffee bean in Northern Areas of Thailand using near infrared spectroscopy *Chiang Mai J. Sci.* **48** 163–75 http://epg.science.cmu.ac.th/ejournal/

[51] Adnan A, Naumann M, Morlein D and Pawelzik E 2020 Reliable discrimination of green coffee beans species: a comparison of UV–vis-based determination of caffeine and chlorogenic acid with non-targeted near-infrared spectroscopy *Foods* **9**

[52] da Silva Araújo C, Macedo L L, Vimercati W C and Saraiva S H 2021 Spectroscopy technique applied to estimate sensory parameters and quantification of total phenolic compounds in coffee *Food Anal. Methods* **14** 1943–52

[53] Baqueta M R, Coqueiro A, Março P H and Valderrama P 2021 Multivariate classification for the direct determination of cup profile in coffee blends via handheld near-infrared spectroscopy *Talanta* **222**

[54] de M, Barbosa S G, dos M B, Scholz S, Kitzberger C S G, de M and Benassi T 2019 Correlation between the composition of green *Arabica coffee* beans and the sensory quality of coffee brews *Food Chem.* **292** 275–80

[55] Fioresi D B, Pereira L L, Catarina da Silva Oliveira E, Moreira T R and Ramos A C 2021 Mid infrared spectroscopy for comparative analysis of fermented arabica and robusta coffee *Food Control* **121**

[56] Brondi A M, Torres C, Garcia J S and Trevisan M G 2017 Differential scanning calorimetry and infrared spectroscopy combined with chemometric analysis to the determination of coffee adulteration by corn *J. Braz. Chem. Soc.* **28** 1308–14

[57] Nogales-Bueno J, Baca-Bocanegra B, Romero-Molina L, Martínez-López A, Rato A E, Heredia F J, Hernández-Hierro J M, Escudero-Gilete M L and González-Miret M L 2020 Control of the extractable content of bioactive compounds in coffee beans by near infrared hyperspectral imaging *LWT* **134**

[58] da C, Araújo S, Macedo L L, Vimercati W C, Ferreira A, Prezotti L C and Saraiva S H 2020 Determination of pH and acidity in green coffee using near-infrared spectroscopy and multivariate regression *J. Sci. Food Agric.* **100** 2488–93

[59] Craig A P, Botelho B G, Oliveira L S and Franca A S 2018 Mid infrared spectroscopy and chemometrics as tools for the classification of roasted coffees by cup quality *Food Chem.* **245** 1052–61

[60] Debona D G, da E C, Oliveira S, Ten Caten C S, Guarçoni R C, Moreira T R, Pereira L L and Moreli A P 2021 Sensory analysis and mid-infrared spectroscopy for discriminating roasted specialty coffees *Coffee Sci.* **16**

[61] Ribeiro J S, de T, Salva J G and Silvarolla M B 2021 Prediction of a wide range of compounds concentration in raw coffee beans using NIRS, PLS and variable selection *Food Control* **125**

[62] Barrios-Rodríguez Y F, Rojas Reyes C A, Triana Campos J S, Girón-Hernández J and Rodríguez-Gamir J 2021 Infrared spectroscopy coupled with chemometrics in coffee post-harvest processes as complement to the sensory analysis *LWT* **145**

[63] Yisak H, Redi-Abshiro M and Chandravanshi B S 2018 Selective determination of caffeine and trigonelline in aqueous extract of green coffee beans by FT-MIR-ATR spectroscopy *Vib. Spectrosc* **97** 33–8

[64] Varão Silva T, Pérez-Rodríguez M, de Oliveira N R, de Santana H and de Almeida L C 2021 Tracing commercial coffee quality by infrared spectroscopy in tandem with pattern recognition approaches *Vib. Spectrosc* **116**

[65] Zhu M, Long Y, Chen Y, Huang Y, Tang L, Gan B, Yu Q and Xie J 2021 Fast determination of lipid and protein content in green coffee beans from different origins using NIR spectroscopy and chemometrics *J. Food Compos. Anal.* **102**

[66] Benes E, Fodor M, Kovács S and Gere A 2020 Application of detrended fluctuation analysis and yield stability index to evaluate near infrared spectra of green and roasted coffee samples *Processes* **8**

[67] USP 2023 *General Information Chemometrics* <1039> United States Pharmacopeia

[68] USP 2023 *Validation of Compendial Procedures* <1225> United States Pharmacopeia

[69] Kementrian Kesehatan Republik Indonesia, e-Farmakope Indonesia VI 2023 (https://efi.kemkes.go.id/webadmin/theories/view/13#side13) (accessed 24 January 2023).

[70] AOAC International Publications 2023 *Appendix K: Guidelines for Dietary Supplements and Botanicals*

Chapter 6

The use of MIR and NIR spectroscopy and chemometrics for the authentication of fruit juice

Abdul Rohman, Anjar Windarsih, Nor Kartini Abu Bakar and Gunawan Indrayanto

Motivated by the beneficial effects of fruit to human health due to the phytochemicals that they contain, fruit juice is one of the most important sectors in the beverage industry. The consumption trends of fruit juice have increased tremendously. However, due to economic reasons, unethical fruit producers may adulterate high-quality fruit juice with lower quality juices. Fruit juices can be adulterated by mixing concentrated pressed juice with cheaper fruit juices, by adding foreign components such as sugars and artificial sweeteners, or by diluting them with water to increase the volume of the fruit juice. The detection of juice adulteration is complex in nature, and therefore accurate and specific methods are required. Infrared spectroscopic based methods, either in near-region (near-infrared, NIR) or in mid-region (mid-infrared, MIR) are promising methods for the authentication of food and beverage products, including fruit juice. Combined with suitable chemometrics techniques of classification and quantification, NIR and MIR spectroscopy have been successfully applied for high value juice, such as grapefruit juice, pomegranate fruit juice, and apple juice. The combination of NIR-MIR spectroscopy and chemometrics is a rapid and green analytical method that is suitable for the routine quality control of juices.

6.1 Introduction

In recent years, the consumer's preference towards nutritious and health beverage commodities has provided a great potential for value-added products such as fruit juices in the global market. Because of their nutritional composition, packaged fruit juices have become one of the most abundantly consumed beverages. The popularity of fruit juices may be attributed to their availability as an affordable and healthy alternative to carbonated beverages. Several juice variants are available in beverage

doi:10.1088/978-0-7503-5573-5ch6 © IOP Publishing Ltd 2023

markets, including minimally processed juices, extra pulp juices, concentrate, pasteurized and unpasteurized juices, and functional juice drinks to meet the consumer's preferences [1].

The production of fruit juice is one of the most important sectors in the beverage industry. According to 'Fruit Juice Market: Global Industry Trends, Share, Size, Growth, Opportunity and Forecast 2021–2026', the market of fruit juice reached a volume of 44.12 billion liters in 2020 and is expected to grow steadily in the coming years [2]. However, due to the economic rewards, unethical fruit producers may adulterate high-quality fruit juice with other components. Fruit juices are typically adulterated by mixing concentrated pressed juice with cheaper fruit juices, by adding sugars and artificial sweeteners, and by diluting them with water to increase the volume of fruit juices. This adulteration practice is carried out by fruit juice manufacturers to increase their profits due to the price difference and to hide low-quality raw materials. The adulteration of fruit juice with other fruit juice adulterations is very common because juice products are complex matrices in nature that are difficult to detect, and therefore sophisticated methods have been proposed and developed for the authentication of fruit juices [3].

Fruit juices that have high prices in the beverage industry, such as grape juice, pomegranate, apple, and Kiwi juice, may potentially be adulterated with lower price juices or adulterated with foreign components, such as sugar, saccharin, etc. For these reasons, instrumental methods based on spectroscopic techniques have been developed, including ultraviolet-visible [4]; atomic absorption and atomic emission, typically by determination of metals in authentic and adulterated fruit juices [5]; NIR and MIR [6], Raman spectroscopy [7], and proton-NMR spectroscopy [8]; and chromatographic-based methods, such as two-dimensional gas chromatography coupled with mass spectrometer (GC x GC-MS) using untargeted volatile metabolomics [9] and high performance liquid chromatography using several detectors to analyse certain components in fruit juices [10]. Due to its property as a fingerprinting analytical technique, infrared spectroscopy, either in near-region (NIR) or mid-region (MIR), may be used as an analytical tool for the authentication of fruit juices, which are typically complex matrices [11].

The responses obtained during the measurements of Fourier transform near-infrared (FT-NIR) and Fourier transform mid-infrared (FT-MIR) spectroscopy are very large. FT-NIR and FT-MIR spectra are characterized by peaks and shoulders at certain wavelengths or wavenumbers, and each peak or shoulder could be characterized by exact wavelength or wavenumbers along with their intensities, which are typically scanned in absorbance mode. It is very challenging to treat data with big responses, and therefore the use of multivariate data analysis or chemometrics is unavoidable [12]. Consequently, almost all publications that apply FT-NIR and FT-MIR spectroscopic techniques employ chemometrics as a complementary data treatment. In the authentication studies of fruit juices, the analytical purposes to be solved are the identification of adulterant types, the classification of authentic and adulterated fruit juices, and the quantification or prediction of adulterant levels in the authentic fruit juices.

6.2 Infrared spectroscopy and chemometrics

Spectroscopy uses the interaction between electromagnetic radiations (EMR) at certain wavelength or wavenumbers with analytes(s) of interest. In relation to infrared (IR) spectroscopy, the interaction between analytes and EMR takes place in the IR region. In analytical applications, the IR region of the EMR spectrum can be divided into three regions, namely far-region (400–10 cm^{-1}), MIR region (4000–400 cm^{-1}), and NIR region (14 285–4000 cm^{-1}). Spectroscopic techniques where the interaction results between analytes and EMR in NIR region and MIR regions are known as NIR and MIR spectroscopy. NIR and MIR spectroscopy can be differentiated from the IR regions used and the type of instrumentation that is used to acquire IR spectra. Other distinctions are based upon the radiation properties, e.g., it is widely agreed that NIR covers EMR with a wavelength range between 750 and 2500 nm [13]. However, some users define NIR at a wavelength range of 650–2500 nm, which overlaps with part of the red-light region of the visible spectrum (400–750 nm). NIR and MIR spectra are generally considered as among the characteristic properties of a compound. Both spectroscopic methods are known as fingerprinting analytical methods. NIR and MIR spectra can be obtained by passing the radiation sources through a sample and determining what fraction of the incident radiation is absorbed at a particular energy or wavelength [14].

Both NIR and MIR spectroscopy are powerful analytical techniques for analyzing inorganic and organic samples, in the form of gases, liquids, and solids samples, either qualitatively or quantitatively. They are based on the concept that atom-atom bonds within molecules vibrate with specific frequencies that may be described by the laws of physics. When these molecular vibrators absorb light with a particular frequency, they are excited to a higher energy level. At room temperature, most molecules are at their rest (zero energy levels). NIR and MIR spectroscopic techniques are suitable for direct analysis of juice samples [15]. NIR and MIR spectra typically involve large data sets, and therefore the use of chemometrics or multivariate data analysis (MDA) is essential.

The term 'chemometrics' was invented by Bruce Kowalski to describe a new way of analyzing chemical data. Chemometrics deals with big data analysis; therefore, chemometrics is also known as MDA in which chemical data coming from instrumental responses were treated and extracted into understandable information. Some definitions of chemometrics can be found in scientific publications and scientific guidelines. In particular, the International Chemometrics Society (ICS) defines chemometrics as: '*The chemical discipline that uses mathematical and statistical methods to design or select optimal measurement procedures and experiments to provide maximum chemical information by analyzing chemical data*' [16].

In analytical applications, the chemometric techniques typically applied in NIR and MIR spectra included preprocessing NIR and MIR spectra, classification (pattern recognition), and multivariate calibrations for quantification [17]. Data preprocessing—including first, second, and third derivatization, mean centering (MC), standard normal variate (SNV), baseline corrections, signal correction and compression, spectra normalizations, smoothing, and multiplicative correction—is often also used to treat NIR spectra and MIR spectra before being subjected to classification chemometrics analysis and multivariate calibrations.

In authentication studies, including juice authentication, the chemometrics of pattern recognition or chemometrics of classification discrimination are the most common. The chemometrics of pattern recognition are typically applied for the classification or discrimination of authentic fruit juices and those adulterated with low-price fruit juices or for the discrimination of authentic fruit juices with those added with foreign components, such as saccharin and sugars [18]. The classification or pattern recognition chemometrics can be performed using three approaches, namely: exploratory data analysis, such as principal component analysis (PCA) and factor analysis (FA); unsupervised pattern recognition, such as hierarchal cluster analysis (HCA) and similarity analysis (SA); and supervised pattern recognition, such as discriminant analysis models and soft independent modeling of class analogy (SIMCA). Discrimination models include linear discriminant analysis (LDA), partial least squares discriminant analysis (PLS-DA), artificial neural networks (ANNs), and k-nearest neighbors (KNNs) [19].

The chemometrics of multivariate calibrations are used to facilitating the quantitative determination of adulterants in authentic fruit juices. Fruit juice is a complex mixture, and therefore the multivariate calibrations modeling the mathematical relationship between the predictor (independent variables) and multiple variables (dependent variables) must be used. Multivariate calibration applies several variables, such as employing the absorbance values at several wavelengths in NIR spectroscopy or wavenumbers region in MIR spectroscopy. The common multivariate calibrations used in the authentication studies include partial least square regression (PLSR), principal component regression (PCR), and stepwise multiple linear regression (SMLR). The accuracy of calibration and validation models using multivariate calibration was evaluated by the coefficient of determination (R^2) for the relationship between two variables, while the precision of models was assessed by the root mean square error of calibration (RMSEC) and root mean square error of prediction (RMSEP). The calibration models should be validated, and the common technique used for the model validation is cross validation. The parameters used for the evaluation of cross validation models are R^2 (for accuracy of the model), root mean square error of cross validation (RMSECV), and the predicted residual error sum of squares (PRESS) [20, 21].

6.3 Authentication of fruit juice using FT-MIR spectroscopy and chemometrics

MIR spectroscopy is mainly related to molecular vibrations. The molecular absorption of EMR in MIR region can cause a transition between the rotational and vibrational energy levels of the ground (lowest) electronic energy state in the molecules. The absorption of MIR radiation is a quantized process, corresponding to the stretching and bending vibrational modes of the bonds in the most covalent molecules. The modes of vibration can either be stretching (in which bond length is observed) or bending (in which bond angle is changed). Stretching can be

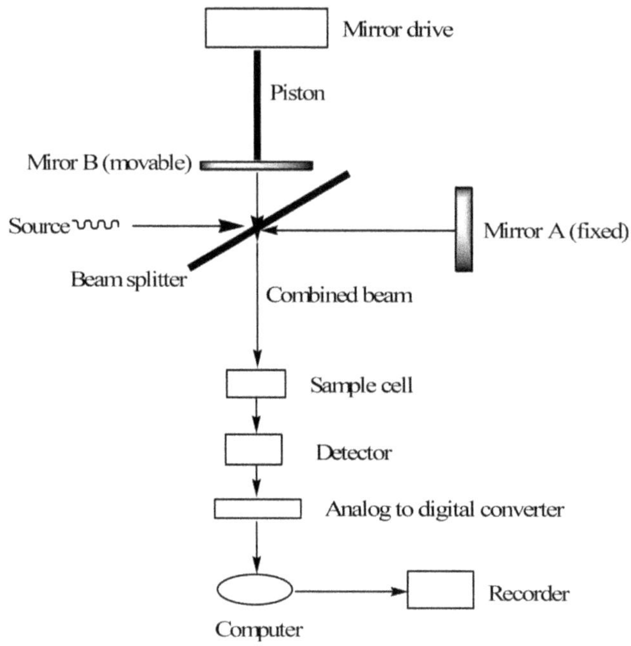

Figure 6.1. Schematic of FTIR spectrometer applying Michelson interferometer. Adapted from [23], Copyright 2013, with permission from Elsevier Ltd.

symmetrical (in plane) or asymmetrical (out of plane). The bending vibration is identified as rock or deformation when moved in the same or in opposite direction, respectively [21].

Dispersive instruments have rarely used in food analyzes due to the difficulties in sample handling technique, and also because they are not equipped with proper spectral acquisition and processing systems in order to give the valuable information for quantitative analysis. Consequently, over the last three decades, FTIR spectroscopy has replaced dispersive IR spectroscopy and has become an important technique for quantitative analysis. FTIR spectroscopy is based on interferometry; therefore, it differs fundamentally from traditional dispersive IR spectroscopy. The Michelson interferometer is employed in most FTIR spectrometers (figure 6.1). An interferometer is composed of two perpendicular mirrors: one is a moving mirror which travels at a constant velocity and the other is a stationary mirror. Between the two mirrors is a beam splitter, normally made up of KBr coated with germanium (for MIR) [22].

The beam splitter divides beam of radiation from the IR source into two parts: one part is reflected to stationary mirror and the other part is transmitted to a moving mirror. When the beams are reflected back, they will recombine at the beam splitter, producing constructive/destructive interference patterns due to the varying difference between the distance traveled by two components of the beam, and part of the recombined beam subsequently reaches the detector. After the IR energy has been selectively absorbed by a sample located between the beam splitter and the detector, the fluctuations in the intensity of energy reaching the detector are digitalized in real time, yielding an interferogram. The interferogram contains all of the information required to produce the IR spectrum of the sample, but this information is in the time domain [24]. To obtain a conventional IR spectrum, the interferogram is further converted into the frequency domain using Fourier transformation, which is a mathematical algorithm used to decode the interferogram and to obtain interpretable information about the individual frequencies in the IR spectrum. Wavelength accuracy is necessary to obtain correct and highly resolved spectra. It is dependent on knowing the exact position of the moving mirror and is achieved using an internal reference laser (He–Ne), which monitors the position of the moving mirror during the scan. This leads to precise scanning and accurate spectrum collection in relation to the wavelength position, which is a key determinant for quantitative spectroscopy.

FT-MIR spectroscopy, in combination with chemometric techniques, is a valuable method for the authentication of high value fruit juices with other fruit juices having low quality and low prices. In addition, this method is also widely reported for the analysis of foreign components, such as sugar, saccharin, water, citric acid etc. Table 6.1 compiles a list of publications reporting the application of FT-MIR spectra and chemometrics for the authentication of fruit juices.

Authentication of pomegranate juice. FTIR spectroscopy in combination with chemometric techniques of PCA and PLSR has been employed for the authentication of pomegranate juice concentrate (PJC) from the adulterants of GJC. The FTIR spectra of both PJC and GJC were compared to search some differences in terms of peaks and shoulders intensity. Using detailed investigation, peak at wavenumbers of 1780–1685 cm^{-1} region corresponding to stretching vibration of carbonyl group (C=O) stretching was finally selected as PCA variables for the differentiation of both juices. PCA using two PCs could clearly separate pure PJC, pure GJC, and PJC samples adulterated with GJC at 2%–14% (vol/vol). Two PCs could explain 99% of the variability in each of these applications. Furthermore, multivariate calibration of PLSR could provide a good relationship between actual values and FTIR predicted values of GJC with R^2 of 0.9751 with low errors. It can be concluded that using appropriate variables of FTIR spectra, the combination of FTIR spectra and chemometrics of PCA and PLSR is suitable for the authentication of PJC from GJC, with accurate and precise classification and prediction results [25]. A recent study has also reported that the combination of FTIR spectroscopy at wavenumbers of 4000–800 cm^{-1} with 128 scannings with resolution of 4 cm^{-1} and chemometrics of PLS-DA and OPLS-DA provide better models that UV–vis spectroscopy for the authentication PJC with dark colored sour cherry and black carrot juices. Before being used for classification modeling, MIR spectra were subjected to some preprocessing including first (FD), second (SD), and third (TD) derivatives, SNV, MSC, along with their combination of FD+SNV, FD+MSC, SD +SNV, SD+MSC, TD+SNV, and TD+MSC. The optimization results show that OPLS-DA model using MIR spectra treated by SNV and MSC has the highest correct classification rate of 100% and 97% for calibration and validation sets,

Table 6.1. Application of FTIR spectroscopy and chemometrics for the authentication of fruit juices.

Issue	FTIR methods	Chemometrics	Results	References
Adulteration of PJC with grape juice concentrate (GJC).	ATR technique, wavenumber range of 4000–650 cm^{-1}, resolution 4 cm^{-1}	PCA and PLS	PCA could differentiate pure PJC and PJC adulterated GJCPLS, and could predict the concentration of GJC in PJC with R^2=0.9751. Lowest level adulterants detected=2%.	[25]
Detection of sugar adulterants (partially inverted cane syrup (PICS), beet sucrose (BS), high fructose corn syrup (HFCS), and a synthetic solution of fructose, glucose, and sucrose (FGS) in apple juices.	ATR technique, wavenumber range of 4000–800 cm^{-1}, resolution 4 cm^{-1}, number of scan 20	PCA and PLS	Prediction results for PICS, BS, HFCS, and FGS achieved an overall correct classification of 96.5%, 93.9%, 92.2%, and 82.4%, respectively.	[26]
Authentication of Concord juice concentration in a grape juice blend.	ATR technique, wavenumber range of 4000–700 cm^{-1}, number of scan 64	SIMCA and PLS	SIMCA demonstrated a high discrimination level of grape juices based on variety. PLS could be applied to predict the percent of Concord grape juice in grape juice with high accuracy and low error of validation (~5%).	[6]
Authentication of sweetness of mango juice: added sugar content (ASC), total soluble solids (TSS), and real juice content (RJC).	ATR technique, wavenumber range of 4000–650 cm^{-1}, scan speed 0.2 cm s^{-1}, number of scan 5	PCA and PLS	PCA could classify mango juices with or without ASC. PLS using wavenumber range of 1476–912 cm^{-1} successfully predicted ASC, TSS, and RJC with R=0.990. Lowest level of adulterants detected=3%–5%.	[27]

(*Continued*)

Table 6.1. *(Continued)*

Issue	FTIR methods	Chemometrics	Results	References
Authentication of pomegranate juice adulterated with sour cherry and black carrot juices.	ATR technique, wavenumber range of 4000–800 cm^{-1}, resolution of 4 cm^{-1}, number of scans 128	PLS-DA, OPLS-DA, and PLS	PLS-DA using transformation of square (seven latent variables) and multiple scattering correction (MSC) (eight latent variables) and OPLS-DA employing SNV and MSC provided the highest correct classification rate of 100% for calibration and 97% for validation. PLS after SNV transformation using eight latent variables demonstrated the best model to predict adulterants (sour cherry and black carrot juices) in pomegranate juice with R^2 validation=0.95.	[28]
Analysis of cane sugar as an added sugar adulterant in apple fruit juices.	ATR technique, wavenumber range of 4000–400 cm^{-1}, resolution of 4 cm^{-1}, number of scans of 32	PCA, SIMCA, and PLS	PCA using wavenumber of 1200–900 cm^{-1} could differentiate authentic and adulterated apple fruit juices with sugar cane. Classification using SIMCA showed 100% accuracy and no misclassification was found. PLS using first derivative spectra from 1145–980 cm^{-1} successfully predicted cane sugar adulterant with R^2=0.991 and low error (RMSE=0.64% v/v) for calibration model and R^2=0.992 with RMSE=0.61% v/v for validation model.	[29]

Detection of corn syrup as an added sweetener in apple fruit juices.	ATR technique, wavenumber range of 4000–400 cm^{-1}, number of scans 32, resolution of 4 cm^{-1}	SIMCA, LDA, and PLS	SIMCA and LDA using variables from 1200–900 cm^{-1} could perfectly discriminate and classify with and without added of corn syrup with 100% correct classification. PLS could be used to detect added corn syrup in apple fruit juices at 1% adulterant concentration with R^2 cross validation=0.9991, RMSE cross validation=0.159% v/v, and relative prediction error of cross validation=1.7% v/v).	[30]
Detection of glucose syrup added in traditional grape, carob, and mulberry pekmez (pekmez=concentrate juices).	ATR technique, wavenumber range of 4000–400 cm^{-1}, resolution of 4 cm^{-1}, number of scans 50	PLS-DA and PLS	PLS-DA successfully discriminated and classified pekmez samples of pure and containing glucose syrup with sensitivity of 100% and efficiency of 100%. PLS regression successfully predicted glucose syrup added in grape, carob, and mulberry pekmez with R^2 calibration of 0.988, 0.998, and 0.996, respectively. The RMSEC was 2.02%, 0.98%, and 1.62% for grape, carob, and mulberry pekmez, respectively. The lowest level of adulterant that could be detected was 2.5%.	[31]

(*Continued*)

Table 6.1. (*Continued*)

Issue	FTIR methods	Chemometrics	Results	References
Detection of adulteration in fruit juices (pineapple, orange, and apple) with grape juice.	ATR technique, wavenumber range of 3070–952 cm^{-1}, resolution of 3.86 cm^{-1}, an optical path length of 20 μm	PCA, HCA, LDA, SVM, and PLS	PCA and HCA demonstrated a clear clustering trend between three types of juices, and between pure and adulterated juices. LDA and SVM could detect the adulteration with an accuracy above 97% for all type of juices. Support vector regression and PLS had high accuracy in predicting grape juice in pineapple, orange, and apple juice with R^2>0.99 and RMSE<1.4% for test set.	[32]
Detection of fraud in lime juice. Samples were collected from markets consisting of two natural classes (Citrus limetta and Citrus aurantifolia) and three reconstructed classes (Spanish concentrate, Chinese concentrate, and concentrate containing adulterants).	KBr technique, wavenumber range of 3950–401 cm^{-1}, resolution of 8 cm^{-1}	PCA, PLS-DA, and counter propagation artificial neural network (CPANN)	PCA could not clearly separate among groups. PLS-DA and CPANN could classify all samples into their groups. CPANN resulted in the best model and the accuracy reached 96% and 87% for the training and validation procedures.	[33]
Differentiation of organic and non-organic grape juices.	ATR technique, wavenumber range of 1550–951 cm^{-1}, resolution of 4 cm^{-1}, number of scans 32	LDA and PLS-DA	Both LDA and PLS-DA could discriminate and classify organic and non-organic grape juices with excellent accuracy rates for both calibration (100%) and validation (100%) models.	[34]

Differentiation of commercial fruit juices (apple, blueberry, blueberry-blended apple, cranberry, cranberry-blended apple, Concord grape, plum, and plum-nectar).	ATR technique, wavenumber range of 4000–700 cm^{-1}, number of scans 128, resolution of 4 cm^{-1}	PCA, HCA, and SIMCA	PCA, HCA, and SIMCA could differentiate all types of commercial fruit juices according to their labels. HCA successfully resulted in natural grouping based on the similarity of their ingredients. SIMCA could be used for the differentiation of fruit juices from different origins without misclassification.	[35]
Detection of orange juice adulteration with glucose, fructose, and sucrose.	ATR technique, wavenumber range of 4000–650 cm^{-1}	Principal component-discriminant function analysis (PC-DFA) and PLS	PC-DFA could differentiate authentic orange juice with those adulterated with glucose, sucrose, and fructose. PLS could be used to detect and predict glucose, fructose, and sucrose in orange juice with RMSE for test samples of 1.7%. The lowest level of adulterants detected was 0.5%–20%.	[36]
Detection of adulteration of freshly squeezed orange juices.	ATR technique, wavenumber range of 4000–800 cm^{-1}, resolution of 4 cm^{-1}, number of scans 32	PCA and LDA	PCA could approximately separate authentic and adulterated juices. LDA could predict adulteration with accuracy of 87.5%. The lowest level adulterants detected 10%.	[37]

respectively. The OPLS-DA model also has high R^2, sensitivity and specificity values with low RMSEE, indicating that good differentiation model was obtained between authentic and adulterated PJC samples [28].

Authentication of apple juice. The combination of FT-MIR spectra and PLSR was successfully applied to determine sugar as an adulterant in apple juice using FTIR spectra at wavenumbers of 4000–400 cm^{-1} at resolution of 4 cm^{-1} with number of scanning of 132. The R^2 values of calibration and prediction for the correlation between actual and FT-MIR predicted values of 0.999 and 0.994, respectively. RMSEC and RMSEP values obtained were 0.005 and 0.018, respectively. These results indicate that the correct selection of peak regions as variables contributes to the accurate and precise determination of sugar in apple juice [38].

Authentication of grapefruit juice. Concord grape juice (CGJ) is considered to be a high-quality juice, which is correlated with many benefits to human health, therefore CGJ can be sold at a premium price. As a consequence, CGJ has the potential to be adulterated with grape juice blends. FT-MIR spectra have the potential to be developed as a rapid method to authenticate CGJ from other grape blends. For this task, MIR spectra combined with SIMCA are used for the classification of 64 CGJ samples as training sets of concentrations of CGJ ranging from 50% to 100%. SIMCA provides a good classification model with SEP of 5.6% using seven factors [6].

6.4 Authentication of fruit juice using FT-NIR spectra and chemometrics

NIR spectroscopy is a nondestructive and simple analytical method that does not require extensive sample preparation. NIR spectroscopy is based on the interaction between analyte and radiation from the NIR region of EMR corresponding to wavelength 700–2500 nm (wavenumbers 12 500–4000 cm^{-1}), which lies between visible light and the MIR region. Some components in samples including fruit juice such as water, carbohydrates, protein, oils and alcohols can absorb NIR radiation. The apparent absorbed EMR by chemical components in the analyzed samples can be calculated from the transmitted or reflected radiation, and the absorbed EMR can be related to the contents of the chemical components [39]. NIR spectra come from the overtones and combination bands of the fundamental vibration of a single chemical bond in the molecules. NIR spectra are composed of many broad and overlapping bands coming from overtones and combination bands from hydrogen-containing groups, namely C–H, N–H, and O–H, which represented the main chemical bonds of organic compounds present in the analyzed samples. In addition, the intensity of NIR spectra also conforms to Lambert–Beer's law. Therefore, the specific chemical bonds could determine the quantity of absorbed light which can be further correlated with the concentrations of analytes(s). Each substance exhibits NIR absorption and has a unique NIR spectrum, therefore the NIR spectrum can be used as a characteristic property of compounds [40]. Combined with chemometrics, NIR spectroscopy has been widely applied in the authentication of food products [41] and quality control of food products [42]. Table 6.2 compiles the application of

Table 6.2. Application of NIR spectroscopy and chemometrics for authentication of fruit juices.

Issue	NIR methods	Chemometrics	Results	References
Detection of lime juice adulteration (adulterants: water and citric acid).	NIR in the spectral range of 900–1700 nm	PCA, PLS-DA, and k-NN	PCA was able to differentiate authentic and adulterated lime juice into two separate clusters. PLS-DA and k-NN using SNV transformation could be used to classify authentic and adulterated lime juice with 88% accuracy.	[43]
Detection of saccharin in commercial fruit juices.	NIR in the wavenumber range of 10 000–4000 cm^{-1}, path length of 0.20 mm, resolution of 2 cm^{-1}	PCA, PLS-DA, PLS	PCA could differentiate fruit juice with those adulterated with saccharin. PLS-DA could differentiate and classify pure fruit juice and adulterated fruit juice with saccharin with R^2=97.90% and RMSE=0.67% weight/volume. PLS could predict and quantify the saccharin added in fruit juice with R^2=97.04%, RMSECV=0.99%, and RMSEP=0.92%.	[3]
Detection of citric acid and tartaric acid in orange juice.	Vis-NIR at spectral range of 325–1075 nm	PLS	PLS could predict the concentration of citric acid in orange juice with R=0.944 and RMSEP=0.596. PLS also successfully detected tartaric acid in orange juice with R=0.930 and RMSEP=0.013.	[45]

(Continued)

Table 6.2. (*Continued*)

Issue	NIR methods	Chemometrics	Results	References
Detection of water adulteration in bayberry juice.	NIR at spectral range of 800–240 nm	PCA, RBFNN (principal component-radial basis function neural networks)	PCA and RBFNN could detect the present of water adulteration in bayberry juice by differentiating authentic and adulterated bayberry juice into two different clusters. The lowest level of adulterations that can be detected by RBFNN was 10% with recognition rate of 97.62%.	[46]
Authentication of grape juices from different varieties.	NIR at spectral range of 400–2500 nm	PCA, LDA, PLS-DA	PCA, LDA, and PLS-DA could be used to differentiate grape juices from different varieties. LDA demonstrated 80% correct classification rate whereas PLS-DA showed 86% correct classification rate.	[47]
Detection of apple juice adulterated with HFCS and sugar solutions.	NIR at spectral range of 400–2498 nm at 2 nm intervals	PLS	PLS regression could detect authentic apple juice with an accuracy of 86%–100% and adulterant apple juice with an accuracy of 91%–100%. The lowest level adulterants that could be detected was 10%.	[48]
Detection of adulterants in fruit juices (apple, pineapple, and orange juices adulterated with grape juice).	NIR at spectral range of 400–2500 nm, resolution of 0.5 nm, path length of 0.5 cm	PCA, LDA, random forest, and support vector regression	PCA could differentiate authentic and adulterated fruit juices. LDA and random forest could discriminate authentic and adulterated fruit juices with an accuracy of 98% in a test set. Support vector regression demonstrated good regression coefficient (0.989) with RMSE of 1.683 in the test set.	[49]

Analysis of sugars in fruit juices (glucose, fructose, and sucrose).	NIR at spectral range of 10 000–4000 cm^{-1} with resolution of 8 cm^{-1}	PLS	PLS could predict the concentration of sugars (glucose, fructose, sucrose) with R^2 of 99.8% and SEP <0.10%.	[50]
Detection of adulteration in apple juice.	NIR at spectral range of 830–2490 nm, resolution of 16 cm^{-1}, number of scans 32	PLS	Data processing using successive projection algorithms (SPA) and particle swarm optimization (PSO) could improve the PLS model. PLS regression demonstrated satisfactory result for detection of adulterants in apple fruit juice with R^2 validation=0.9986 and RMSEP=0.0628.	[51]
Determination of sucrose, glucose, and fructose in sweet sorghum juice.	NIR at spectral range of 10 000–4000 cm^{-1} with resolution of 4 cm^{-1} and number of scans 32	PLS	PLS regression could quantify the level of sucrose, glucose, and fructose with high accuracy and low error (RMSEP of 4, 1, and 0,6 mg ml^{-1}, respectively).	[52]
Detection of Monilia contamination in plum juice.	NIR at spectral range of 900–1700 nm	LDA and PLS	LDA could differentiate pure plum juices and contaminated plum juices in a clear result. PLS regression could predict the Monilia contamination in plum juices with R^2 of 0.87 and 0.78 and RMSE of 3.98% and 5.09% for calibration and validation model, respectively.	[53]

(*Continued*)

Table 6.2. (*Continued*)

Issue	NIR methods	Chemometrics	Results	References
Determination of sugars (glucose, fructose, maltose, sucrose, and lactose) and carbohydrate in fresh juices (apple, banana, cucumber, celery, orange, and melon).	NIR at spectral range of 740–1070 nm, spectral resolution was 1 nm and 0.6 nm	PLS	PLS regression could determine the sugars content with R^2 validation of 0.89, RMSEP of 0.95%, and RPD (ratio of performance to deviation) of 3.06. For carbohydrate determination, PLS regression could determine with R^2 validation of 0.96, RMSEP of 0.63%, and RPD of 5.06.	[54]

NIR spectroscopy in combination with chemometrics for the authentication of fruit juice.

NIR spectra coupled with chemometrics of classification of PLS-DA and multivariate calibration of PLSR have been employed for the authentication of commercial fruit juice samples from the adulterant of saccharin. In this study, six different fruit juice samples that are commercially available in the market were intentionally mixed with saccharin (as adulterant) with 10 different percentage levels of saccharin at 0.10%–2.00% (weight/volume). A total of 198 samples consisting of 18 pure juice samples (unadulterated) and 180 adulterated juice samples with saccharin was used in calibration model. PLS-DA model was carried out for the differentiation between unadulterated (authentic) fruit juice samples and those adulterated with saccharin. The PLS-DA model provides R^2 value of 97.90% with an RMSE error of 0.67% indicating accurate classification models for two classes (authentic and adulterated fruit juices with saccharin). Furthermore, PLSR was successfully used to predict saccharin levels in juice samples resulting of R^2 value for the correlation between actual values and FTIR predicted values of 97.04 with RMSECV of 0.88%. The PLSR calibration model was then subjected to cross validation using the leave one out technique for 30% of the total adulterated juice samples (60 adulterated fruit juice samples). The excellent performance of the model was obtained with low RMSEP value of 0.92% and high correlation factor of 0.97, indicating that the developed method is accurate, robust, and precise for authentication of fruit juices from saccharin adulterant [3].

Handled-NIR spectrophotometer at wavelength 900–1700 nm in combination with chemometrics and machine learning has been proposed and validated for the authentication analysis of lime fruit juice from the addition of adulterant of water and citric acid. PCA could provide two clusters of authentic lime fruit juice and that adulterated with water-citric acid. PLS-DA and k-nearest neighbor's (k-NN) algorithms with different spectral preprocessing techniques including MSC, SNV, and derivative (first and second derivate spectra) were applied as predictive models. Based on the optimization results, PLS-DA with SNV transforming offered the most accurate prediction for the adulteration model of lime fruit juice with water-citric acid. PLS-DA model was capable of classifying juices with an accuracy of 88% and a Matthew's r-value of 0.75 in the external validation set. In addition, the k-NN model with SNV transforming had the highest accuracy and Matthew's r-value of 88% and 0.76, respectively. The results indicate that a handled-NIR spectrophotometer in combination with chemometrics provides a promising tool for rapid detection of lime juice adulterated with foreign components (water and citric acid). The model needs to be applied for analysis of lime fruit juice with a wider range of variability to increase the robustness of the analytical method [43]. NIR spectra was also successful for the authenticity of fruit juices, i.e., orange, apple, and grapefruit juices, from the adulterant of water and sugar by determining some parameters contributing to the juice's quality including soluble dry matter, specific density, total acids, lactic acid, ascorbic acid, malic acid, citric acid, and ethanol content. Fruit juices that have lost one of these parameters are considered to be unauthentic. Combined with a three-dimensional display PCA, NIR spectra could separate these authentic fruit juices from those with added water and sugar solution with adulterant levels as low as 1% [44].

References

[1] Rai P, Mehrotra S and Sharma S K 2022 Challenges in assessing the quality of fruit juices: intervening role of biosensors *Food Chem.* **386** 132825

[2] Calle L P, Barea-sep M, Ruiz-rodr A, Ferreiro-gonz Á Á, Palma M and M 2022 Juices using machine learning tools and spectroscopy data *Sensors* 1–12

[3] Mabood F, Hussain J, Jabeen F, Abbas G, Allaham B and Albroumi M *et al* 2018 Applications of FT-NIRS combined with PLS multivariate methods for the detection and quantification of saccharin adulteration in commercial fruit juices *Food Addit. Contam.—Part A Chem. Anal. Control Expo. Risk. Assess* **35** 1052–60

[4] Boggia R, Casolino M C, Hysenaj V, Oliveri P and Zunin P 2013 A screening method based on UV-Visible spectroscopy and multivariate analysis to assess addition of filler juices and water to pomegranate juices *Food Chem.* **140** 735–41

[5] Anastácio M, dos Santos A P M, Aschner M and Mateus L 2018 Determination of trace metals in fruit juices in the Portuguese market *Toxicol. Rep.* **5** 434–9

[6] Snyder A B, Sweeney C F, Rodriguez-Saona L E and Giusti M M 2014 Rapid authentication of concord juice concentration in a grape juice blend using Fourier-transform infrared spectroscopy and chemometric analysis *Food Chem.* **147** 295–301

[7] Mizrach A, Schmilovitch Z, Korotic R, Irudayaraj J and Shapira R 2007 Yeast detection in apple juice using Raman spectroscopy and chemometric methods *Trans. ASABE* **50** 2143–9

[8] Vigneau E and Thomas F 2012 Model calibration and feature selection for orange juice authentication by 1H NMR spectroscopy *Chemom. Intell. Lab. Syst.* **117** 22–30

[9] Li S, Hu Y, Liu W, Chen Y, Wang F and Lu X *et al* 2020 Untargeted volatile metabolomics using comprehensive two-dimensional gas chromatography-mass spectrometry—a solution for orange juice authentication *Talanta* **217** 121038

[10] Yeganeh-Zare S, Farhadi K and Amiri S 2022 Rapid detection of apple juice concentrate adulteration with date concentrate, fructose and glucose syrup using HPLC-RID incorporated with chemometric tools *Food Chem.* **370** 131015

[11] Sanchez P D C, Arogancia H B T, Boyles K M, Pontillo A J B and Ali M M 2022 Emerging nondestructive techniques for the quality and safety evaluation of pork and beef: recent advances, challenges, and future perspectives *Appl. Food Res.* **2** 100147

[12] Bureau S, Cozzolino D and Clark C J 2019 Contributions of Fourier-transform mid infrared (FT-MIR) spectroscopy to the study of fruit and vegetables: a review *Postharvest Biol. Technol.* **148** 1–14

[13] Guillén M D and Cabo N 1997 Characterization of edible oils and lard by fourier transform infrared spectroscopy. Relationships between composition and frequency of concrete bands in the fingerprint region *J. Am. Oil Chem. Soc.* **74** 1281–6

[14] Rohman A, Windarsih A, Hossain M A M, Johan M R, Ali M E and Fadzilah N A 2019 Application of near- and mid-infrared spectroscopy combined with chemometrics for discrimination and authentication of herbal products: a review *J. Appl. Pharm. Sci.* **9**

[15] Jamwal R, Amit , Kumari S, Sharma S, Kelly S and Cannavan A *et al* 2021 Recent trends in the use of FTIR spectroscopy integrated with chemometrics for the detection of edible oil adulteration *Vib. Spectrosc.* **113** 103222

[16] Rohman A, Windarsih A, Riyanto S, Sudjadi , Shuhel Ahmad S A and Rosman A S *et al* 2016 Fourier transform infrared spectroscopy combined with multivariate calibrations for the authentication of avocado oil *Int. J. Food Prop.* **19** 680–7
[17] Moros J, Garrigues S and Guardia M de la 2010 Vibrational spectroscopy provides a green tool for multi-component analysis *TrAC—Trends Anal. Chem.* **29** 578–91
[18] Kusumadewi A P, Martien R, Pramono S, Setyawan A A and Rohman A 2022 Review on the application of chemometrics for the standardization and authentication of *Curcuma xanthorrhiza Food Res.* **6** 1–8
[19] Esteki M, Shahsavari Z and Simal-Gandara J 2018 Use of spectroscopic methods in combination with linear discriminant analysis for authentication of food products *Food Control* **91** 100–12
[20] Yizeng L, Hailong W U, Guoli S, Jianhui J, Sheng L and Ruqin Y U 2006 Aspects of recent developments in analytical chemometrics **49** 193–203
[21] Rohman A, Ghazali M A B, Windarsih A, Irnawati , Riyanto S and Yusof F M *et al* 2020 Comprehensive review on application of FTIR spectroscopy coupled with chemometrics for authentication analysis of fats and oils in the food products *Molecules* **25** 1–28
[22] Karoui R, Downey G and Blecker C 2010 Mid-infrared spectroscopy coupled with chemometrics: a tool for the analysis of intact food systems and the exploration of their molecular structure-quality relationships-a review *Chem. Rev.* **110** 6144–68
[23] Silverstein R M, Francis X and Webster D J K 2005 Spectrometric identification of organic compounds *J. Mol. Struct* 512 http://linkinghub.elsevier.com/retrieve/pii/002228607687024X
[24] Rohman A 2019 The employment of Fourier transform infrared spectroscopy coupled with chemometrics techniques for traceability and authentication of meat and meat products *J. Adv. Vet. Anim. Res.* **6**
[25] Vardin H, Tay A, Ozen B and Mauer L 2008 Authentication of pomegranate juice concentrate using FTIR spectroscopy and chemometrics *Food Chem.* **108** 742–8
[26] Kelly J F D and Downey G 2005 Detection of sugar adulterants in apple juice using fourier transform infrared spectroscopy and chemometrics *J. Agric. Food Chem.* **53** 3281–6
[27] Jha S N and Gunasekaran S 2010 Authentication of sweetness of mango juice using Fourier transform infrared-attenuated total reflection spectroscopy *J. Food Eng.* **101** 337–42
[28] Aykac B, Cavdaroglu C and Ozen B 2023 Authentication of pomegranate juice in binary and ternary mixtures with spectroscopic methods *J. Food Compos. Anal.* **117**
[29] Dhaulaniya A S, Balan B, Yadav A, Jamwal R, Kelly S and Cannavan A *et al* 2020 Development of an FTIR based chemometric model for the qualitative and quantitative evaluation of cane sugar as an added sugar adulterant in apple fruit juices**37** 539–51
[30] Dhaulaniya A S, Balan B, Sodhi K K, Kelly S, Cannavan A and Singh D K 2020 Qualitative and quantitative evaluation of corn syrup as a potential added sweetener in apple fruit juices using mid-infrared spectroscopy assisted chemometric modeling *LWT* **131** 109749
[31] Yaman N and Velioglu S D 2019 Use of attenuated total reflectance—Fourier transform infrared (ATR-FTIR) spectroscopy in combination with multivariate methods for the rapid determination of the adulteration of grape, carob and mulberry PEKmez *Foods* **8** http://www.ncbi.nlm.nih.gov/pubmed/31261701
[32] Calle J L P, Ferreiro-González M, Ruiz-Rodríguez A, Fernández D and Palma M 2022 Detection of adulterations in fruit juices using machine learning methods over FT-IR spectroscopic data *Agronomy.* **12** 683

[33] Mohammadian A, Barzegar M and Mani-Varnosfaderani A 2021 Detection of fraud in lime juice using pattern recognition techniques and FT-IR spectroscopy *Food Sci. Nutr.* **9** 3026–38

[34] Junges C H, Guerra C C, Gomes A A and Ferrão M F 2022 Green analytical methodology for grape juice classification using FTIR spectroscopy combined with chemometrics *Talanta Open* **6** 100168

[35] He J, Rodriguez-Saona L E and Giusti M M 2007 Midinfrared spectroscopy for juice authentication-rapid differentiation of commercial juices *J. Agric. Food Chem.* **55** 4443–52

[36] Ellis D I, Ellis J, Muhamadali H, Xu Y, Horn A B and Goodacre R 2016 Rapid, high-throughput, and quantitative determination of orange juice adulteration by Fourier-transform infrared spectroscopy *Anal. Methods* **8** 5581–6

[37] Shen F, Wu Q, Su A, Tang P, Shao X and Liu B 2016 Detection of adulteration in freshly squeezed orange juice by electronic nose and infrared spectroscopy *Czech J. Food Sci.* **34** 224–32

[38] Singh S and Thakur R 2016 Sugar analysis in apple juice by FTIR spectroscopy with PLS regression *IJSTE -Int. J. Sci. Technol. Eng.* **2** 413–5

[39] Esteve Agelet L and Hurburgh C R 2014 Limitations and current applications of Near Infrared Spectroscopy for single seed analysis *Talanta* **121** 288–99

[40] Li X, Zhang L, Zhang Y, Wang D, Wang X and Yu L *et al* 2020 Review of NIR spectroscopy methods for nondestructive quality analysis of oilseeds and edible oils *Trends Food Sci. Technol.* **101** 172–81

[41] Nobari Moghaddam H, Tamiji Z, Akbari Lakeh M, Khoshayand M R and Haji Mahmoodi M 2022 Multivariate analysis of food fraud: a review of NIR based instruments in tandem with chemometrics *J. Food Compos. Anal.* **107** 104343.

[42] Porep J U, Kammerer D R and Carle R 2015 On-line application of near infrared (NIR) spectroscopy in food production *Trends Food Sci. Technol.* **46** 211–30

[43] Jahani R, Yazdanpanah H, van Ruth S M, Kobarfard F, Alewijn M and Mahboubi A *et al* 2020 Novel application of near-infrared spectroscopy and chemometrics approach for detection of lime juice adulteration *Iran J. Pharm. Res.* **19** 34–44

[44] Šnurkovic P 2013 Quality assessment of fruit juices by NIR spectroscopy *Acta Univ. Agric. Silvic. Mendelianae Brun.* **61** 803–12

[45] Cen H, Bao Y, He Y and Sun D W 2007 Visible and near infrared spectroscopy for rapid detection of citric and tartaric acids in orange juice *J. Food Eng.* **82** 253–60

[46] Xie L J, Ye X Q, Liu D H and Bin Y Y 2008 Application of principal component-radial basis function neural networks (PC-RBFNN) for the detection of water-adulterated bayberry juice by near-infrared spectroscopy *J. Zhejiang Univ. Sci. B.* **9** 982–9

[47] Cozzolino D, Cynkar W, Shah N and Smith P 2012 Varietal differentiation of grape juice based on the analysis of near- and mid-infrared spectral data *Food Anal. Methods* **5** 381–7

[48] León L, Daniel Kelly J and Downey G 2005 Detection of apple juice adulteration using near-infrared transflectance spectroscopy *Appl. Spectrosc.* **59** 593–9

[49] Calle J L P, Barea-Sepúlveda M, Ruiz-Rodríguez A, Álvarez J Á, Ferreiro-González M and Palma M 2022 Rapid detection and quantification of adulterants in fruit juices using machine learning tools and spectroscopy data *Sensors* **22** 3852

[50] Rodriguez-Saona L E, Fry F S, McLaughlin M A and Calvey E M 2001 Rapid analysis of sugars in fruit juices by FT-NIR spectroscopy *Carbohydr. Res.* **336** 63–74

[51] Li Y, Guo Y, Liu C, Wang W, Rao P and Fu C *et al* 2017 SPA combined with swarm intelligence optimization algorithms for wavelength variable selection to rapidly discriminate the adulteration of apple juice *Food Anal. Methods* **10** 1965–71
[52] Simeone M L F, Parrella R A C, Schaffert R E, Damasceno C M B, Leal M C B and Pasquini C 2017 Near infrared spectroscopy determination of sucrose, glucose and fructose in sweet sorghum juice *Microchem. J.* **134** 125–30
[53] Vitalis F, Nugraha D T, Aouadi B, Bósquez J P A, Bodor Z and Zaukuu J L Z *et al* 2021 Detection of monilia contamination in plum and plum juice with NIR spectroscopy and electronic tongue *Chemosens 2021* **9** 355
[54] Chen W, Li H, Zhang F, Xiao W, Zhang R and Chen Z *et al* 2021 Handheld short-wavelength NIR spectroscopy for rapid determination of sugars and carbohydrate in fresh juice with sampling error profile analysis *Infrared Phys. Technol.* **115** 103732

Chapter 7

Advances in the analysis of fruits by near-infrared spectroscopy

D Cozzolino and B Dayananda

The evaluation of the proximate composition, functional, nutritional, and health properties of fruits and fruit products faces many challenges. Functional and nutritive properties of fruits are analyzed using a wide range of advanced instrumental methods and techniques. In addition to the classic techniques that are used to measure or characterize the chemical composition of fruits, such as chromatography (e.g. gas chromatography), high performance liquid chromatography (HPLC), mass spectrometry (MS), and electrophoresis, advances in molecular spectroscopy, e.g. near-infrared (NIR) and mid-infrared (MIR) spectroscopy, Raman, and hyperspectral imaging, have been implemented and reported by several authors. Among the molecular spectroscopic techniques, NIR spectroscopy has been the most widely used. This chapter presents and discusses applications and developments of NIR spectroscopy combined with data analytics or chemometric tools to assess the composition of different fruits along the supply and value chains (e.g. maturity, chemical composition, and authenticity).

7.1 Introduction

The evaluation of the proximate composition, functional, nutritional, and health properties of fruits and fruit products faces many challenges [1–5]. These challenges can be associated with several issues, including how to ensure that fruits meet mandatory and minimal quality control standards; how changes during process and manufacturing might lead to changes in the quality and nutritional value of the fruit, and how these changes can be identified; how to monitor and detect adulteration, origin (e.g. variety identification) or fraud (e.g. organic vs non-organic production); and how to identify the geographical origin or traceability of the fruits, as well as their derived products along the supply value chain [1–5]. Many of these issues have been addressed by the current advances in instrumental methods and techniques, as

doi:10.1088/978-0-7503-5573-5ch7 © IOP Publishing Ltd 2023

well as data analytics, which include chromatography (e.g., gas chromatography, GC), HPLC, MS, electrophoresis, and other sensitive as well as specific analytical tools. In addition, advances in molecular spectroscopy such as NIR and MIR spectroscopy, Raman spectroscopy, and hyperspectral image techniques have enabled several samples to be handled in an industrial manufacturing setting [6–13].

The nutritional value of fruits and vegetables is high, and they contain several macronutrients and bioactive compounds of importance to human health [14–17]. Some of these macronutrients are commonly found in fruits, vegetables, and other edible plants, including carbohydrates (e.g. starch, sugars), protein, and lipids (e.g. saturated and unsaturated fatty acids) [14–17]. In addition, fruits and vegetables contain micronutrients (e.g., minerals and vitamins), bioactive ingredients (e.g., phytochemicals), and dietary fiber that can prevent chronic diseases, cancer, and other non-communicable diseases [14–17]. Fruits and vegetables are also defined as functional foods due to their protective effects against human diseases [14–17]. For example, phenolic compounds, especially flavonoids, are important anti-cancer molecules [14–17]. These bioactive compounds are also involved in the first defense line mechanism against oxidative stress via the quenching of a singlet oxygen in the human body [14–17]. The other bioactive or phytochemicals present in fruits and vegetables are important because they play a role in the visual properties of the sample, where different pigments are responsible for the color and appearance of the fruit [14–17]. For instance, green color is mostly derived from the presence of glucosinolates and chlorophylls; anthocyanins give the colors red, orange, blue, and purple, depending on the pH; while carotenoids are responsible for yellow, orange, and red colors found in fruits and vegetables [18–21].

Due to the large quantity of information recorded during the utilization of analytical instrumental techniques, the incorporation of data analytics and chemometrics methods has become essential for the analysis and interpretation of the information into useful data, playing an important role in the development of a given application [22–25]. The so-called classic instrumental methods are used and applied to a vast number and type of samples in the field of fruit and vegetable analysis, where these technologies usually require several steps prior to or during analysis, such as separation and fractionation of the sample prior analysis [22–25]. Rapid instrumental methods based on molecular spectroscopy, such as NIR spectroscopy, have been explored to analyze the composition, nutritional value, and functional properties of fruits and vegetables [8–10, 24–28]. The application of NIR spectroscopy has allowed for the definition of the so-called fingerprinting of a sample, or even to identify single compounds [8–10, 24–28]. The utilization of this technique has allowed for the development and implementation of rapid and high throughput methods, which require minimal sample preparation (e.g. not chromatographic separation) and which can be used in any step along the supply and value chains of fruits [8–10, 24–29].

The use of NIR spectroscopy as an analytical tool was not considered until the 1960s due to the broad absorbance bands that characterize its spectrum [30–33]. Many applications have been developed over the last 30 years following the development of electronics and computers, concomitantly with the relative simplicity of using NIR instruments (e.g. portable, handheld) [30–35]. The development and implementation of such applications have also been made possible by developments in chemometrics [36–42]. By the 1980s, the use of chemometrics became more popular among scientists and analysts because computing power enabled the control of instruments and the development of models [36–42]. Methods such as PCA, and predictive methods such as principal component regression (PCR) and partial least squares (PLS), are now widely used in the development of NIR applications to predict the composition and quality of a wide range of fruits and vegetables [36–42]. These methods have gained attention due to their ability to provide rapid, non-destructive, and precise information of compounds and other properties in these types of samples [36–42].

This chapter presents and discusses applications as well as developments of NIR spectroscopy combined with data analytics or chemometric tools to assess the composition and quality of different fruits along the supply and value chains (e.g. maturity, chemical composition, authenticity).

7.2 Origin and characteristics of near-infrared spectroscopy

Molecular spectroscopy-based methods such as NIR and MIR spectroscopy, Raman spectroscopy, and in recent years the incorporation of hyperspectral imaging have been explored as analytical methods to analyze fruits and vegetables due to the many advantages that they can offer [43–46]. Their advantages include the possibility of non-destructive analysis, minimal or no sample preparation is required, and they do not use reagents [43–46]. In addition, recoding the spectra of a sample takes a few seconds and several properties can be estimated from a single analysis. NIR spectroscopy is based on the overtones and vibrations of the atoms present in a molecule when passing infrared radiation through a tested sample [43–46]. In the NIR region, the bonds of most of the organic molecules vibrate at different frequencies, including those generated from C–H, O–H, N–H, C=O, and other functional groups determining the spectrum of a sample [43–46]. Different bio (e.g. polyphenols) and chemical compounds (e.g. sugars) present in fruits can be determined by the amount and frequency of absorbed light and the quantity of reflected or transmitted light [43–46]. The molecular bonds present in the fruit matrix vibrate at specific frequencies, which are determined by the mass of the constituent atoms, the shape of the molecule, the stiffness of the bonds, and the periods of the associated vibrational coupling [43–46]. The NIR spectra of a fruit result from complex overtones and high frequency combinations at shorter wavelengths. Spectral peaks in the MIR frequencies are often sharper and better resolved than in the NIR domain. The higher overtones of the O–H (oxygen–hydrogen), N–H (nitrogen–hydrogen), C–H (carbon–hydrogen), and S–H (sulfur–hydrogen) bands from the MIR wavelengths are still observed in the NIR region, although they are much weaker than the fundamental frequencies in the MIR [43–46]. In addition to the existence of combination bands (e.g. C–O stretch and N–H bend in protein), they also give rise to a crowded NIR spectrum with strongly overlapping bands [43–46].

A major disadvantage of this characteristic overlap and complex NIR spectra has been the difficulty of quantification and interpretation of data by direct observation of the NIR spectra [43–46]. In recent years, new instrumentation combined with different algorithms have taken advantage of this complexity, and have made the technique much more powerful and simple to use [43–46]. More complex molecules have many bonds; vibrations that can be also conjugated, leading to two possible modes of vibration: stretching and bending [43–46].

7.3 Data analysis and chemometrics

Chemometrics is defined as '*the art of extracting chemically relevant information from data produced in chemical experiments or instrumental techniques*' [37–39, 46–48]. The application of chemometrics and other techniques, such as pattern recognition and design of experiments, is closely connected with the application and utilization of NIR spectroscopy, and has gained more scientific credibility [37–39, 46–48]. The scope of chemometrics is wide, and applications of these techniques are found in many fields where the toolbox of useful methods is diverse. As described earlier, the spectral information recorded originates from the occurrence of band positions, intensities, and shapes [37–39, 46–48]. Chemometrics provide a means to move beyond the one-dimensional (univariate) space. It can also be very helpful to reveal patterns and trends in data sets that are important due to the various interferences and interactions in the matrix [37–39, 46–48]. Today, many modern instrumental measurement techniques are multivariate and based on indirect measurements of the chemical and physical properties of a sample [37–39, 46–48]. The use of NIR spectroscopy provides the possibility to obtain more information from a single measurement because they can record the responses of many variables (absorbance values at different wavelengths) simultaneously, which makes it essential to use chemometrics to extract this information [37–39, 46–48]. Although these powerful methods combined with the computer technology necessary to use them have only become readily available in recent years, their use has significantly boosted the development of instrumental applications, including NIR spectroscopy [37–39, 46–48].

A broad range of techniques are now available, including data reduction, regression and classification methods, and preprocessing techniques. Principal component analysis (PCA) is a commonly used data compression and visualization tool [37–39, 46–48]. Other techniques are available for sample classification or discrimination, e.g., soft independent modeling of class analogy, linear discriminant analysis (LDA), hierarchical cluster analysis, FDA, ANN, and PLS [37–39, 46–48]. The application of artificial neural networks (ANNs) is a more recent technique for data and knowledge processing that is characterized by its analogy with a biological neuron [28–35]. When the firing frequency of a neuron is compared with that of a computer, then for a neuron this frequency is rather low [37–39, 46–48]. In the biological neuron, the input signal from the dendrites travels through the axons to the synapse. There the information is transformed and sent across the synapse to the dendrites of the next neuron [37–39, 46–48]. The use of ANN is also being applied to

NIR and is incorporated into the software provided by some instrument companies [37–39, 46–48]. ANN allows the program to learn to recognize certain spectral features by comparison to a learning set. ANN can also improve calibrations, which has enabled the use of infrared (particularly in NIR applications dealing with nonlinear issues) as a fingerprinting technique. Other methods such as PLS and PCR, have been extensively used to develop correlative models between the NIR spectra and the reference data, e.g. sugar concentration, moisture or dry matter (DM) [37–39, 46–48]. These models (e.g. calibrations) are evaluated using statistical techniques such as the coefficient of determination (calibration and prediction) (R^2), the standard error in cross validation or prediction (SEP), and the residual predictive value (RPD) [49]. Preprocessing techniques as well as calibration transfer from one spectrophotometer to another (e.g. calibration transfer) have been made possible through the utilization of chemometrics [37–39, 46–48].

7.3.1 Application of NIR spectroscopy in fruit analysis

The applications and utilization of NIR spectroscopy in fruits to analyze and monitor composition are vast [50, 51]. A number of reports have demonstrated, compared, and discussed the use of different instruments (e.g. portable, bench, fiber optics), measurement principles (e.g. transmission, reflectance), novel techniques (e.g. time and spatially resolved spectroscopy for the estimation of light absorption and scattering properties of vegetable tissue), as well as the combination of NIR with multi- and hyperspectral imaging techniques to measure fruit composition [50, 51]. Applications of NIR spectroscopy have been focused on the non-destructive measurement of soluble solids (TSSs) content in different fruit species and varieties [50, 51]. The use of visible (VIS) and NIR spectroscopy, as well as multispectral imaging and hyperspectral imaging have also been evaluated to measure composition (e.g. sugars, acidity) and quality attributes (e.g firmness, diseases), as well as the identification of fruit species (e.g. apple, orange, kiwifruit, peach, grape, strawberry, grape, jujube, banana, mango, and others), authenticity, and production system (e.g. organic vs non-organic) [50, 51].

7.3.2 Maturity

NIR spectroscopy is typically used to measure or estimate the maturity of fruits [50, 51]. Scientists have evaluated a wide range of fruit maturity indicators, including the use of chronological, physical, and biochemical measurements or parameters, most of them destructive [50, 51]. Rapid and non-destructive techniques in combination with chemometrics have been developed by several research groups to monitor and predict fruit maturity at harvest in different species and varieties [50, 51].

Among these methods and techniques, NIR spectroscopy has been shown to be the most accurate and versatile to be used as tool to estimate fruit maturity at harvest [50, 51]. The spectral data obtained in addition to other parameters have been combined to develop specific calibration models based on NIR spectroscopy [50, 51]. Different chemometrics or algorithms have been also evaluated and applied to estimate fruit maturity [50, 51]. However, models developed using a specific variety may not necessarily be directly applicable to other varieties of the same fruit species [52]. The use of a pre-existing calibration model developed using the Kensington Pride mango data set installed in a VIS-NIR (Felix F-750) spectrophotomer to predict several mango varieties not included in the original data set has been studied [52]. The results of this study indicated that the pre-calibrated models were not adequate to predict the composition of local mango varieties and recalibration was necessary [52]. Prediction models developed using PLS regression were adequate to predict maturity parameters in apple, mango, grapes, melon, and kiwifruit, while nonlinear models such as least squares support vector machine were better to predict maturity in pear and peach samples [52].

NIR spectroscopy has also been used to develop calibration models to predict TSS content in navel orange fruits [53]. The spectra of oranges collected using a VIS-NIR (350–1800 nm) instrument was preprocessed using multiplicative scatter correction (MSC) and standard normal variate correction (SNV) techniques [53]. Calibration models for the prediction of TSS content were developed using PLS regression and BPNN based on PCA [53]. The PCA-BPNN models developed using MSC yielded the best predictive results R^2 of 0.90 and root mean square error of prediction (RMSEP) of 0.68 °Brix [53]. The results of this study indicated that PCA-BPNN can be considered to be a suitable tool to deal with the nonlinearities in the model, showing additional advantages over PLS regression to measure TSS content in orange samples [53].

7.3.3 Sugars

In parallel with the estimation of maturity, the measurement of sugars has been one of the classical applications of NIR spectroscopy in the fruit value chain, as reported by different authors [50, 51, 54]. The use of Fourier transform NIR (FT-NIR) spectroscopy combined with chemometrics was reported to measure sugars in different fruit juice samples [54]. In this study, different sample preparation methods were evaluated by creating different aqueous solutions of sugar mixtures (glucose, fructose, and sucrose; 0%–8% w/v) that were utilized to build a PLS regression model [54]. The authors of this study evaluated the use of reflectance and transmittance (using a 0.5 mm cell), as well as reflectance using a fiberglass paper filter to collect the spectra of the fruit samples [54]. Calibration models were developed using PLS regression with cross-validation (leave-one-out approach), where two reference methods were also tested, i.e., HPLC and standard enzymatic techniques [54]. The PLS regression models developed using samples scanned in transmittance provided the best performance with a SEP higher than 0.10% and R^2 of 0.99, which accurately and precisely predicted the content of sugar in the juice samples analyzed, whereas poor calibration statistics were reported using samples scanned in reflectance [54]. Overall, these researchers concluded that the use of FT-NIR spectroscopy allowed for the fast and accurate analysis of sugars in fruit juice samples [54].

The ability of NIR spectroscopy to measure the chemical composition (e.g. moisture, total sugars, acidity, TSS, pH, and ascorbic acid) of guava and passion fruit samples has been evaluated [55]. Fruit samples were analyzed using both NIR spectroscopy and reference methods, where PLS regression was used to develop the calibration models and tested using the root mean squares error of calibration (RMSEC), RMSEP and R^2 [55]. The PLS calibration models for moisture content and total sugars yield a RMSEP of 0.24 and 0.27, respectively, for guava pulp; and RMSEP of 0.40 and 0.41, respectively, for passion fruit pulp [55]. A comparison of reflectance, interactance, and transmission in the NIR region between 500 and 1100 nm on the robustness of the PLS regression models developed to measure TSS and titratable acidity in Satsuma Mandarin (*Citrus reticulata* cv. 'Miyagawa') has been reported [56]. The use of transmission in the NIR region between 700 and 930 nm delivered the best model for the measurement of TSS content (R^2:0.93, RMSEP; 0.32%) [56]. However, no calibrations were obtained for the prediction of TA ($R^2 <$ 0.65, RMSEP = 0.15%) [56]. The evaluation of two portable NIR devices (VIS-NIR and AOTF-NIR spectrophotometers) to measure the composition of oranges non-destructively has also been reported [57]. In this study, TSS, acidity, TA, maturity index, flesh firmness, juice volume, fruit weight, rind weight, juice volume to fruit weight ratio, fruit color index, and juice color index were measured and predicted using NIR spectroscopy [57]. The RPD values obtained for the prediction of these parameters ranged from 1.67 to 2.21 with the VIS-NIR instrument and between 1.03 and 2.33 using the AOTF instrument [57].

The use of transmission and reflectance were compared to measure TSS content, pH, and firmness in pomegranate samples [58]. The calibration models for TSS content developed using PLS regression and the NIR spectra collected in transmission gave a *R*: 0.95, RMSEC: 0.22 °Brix and RPD: 6.7, while the validation statistics reported delivered a *R*: 0.94, RMSEP: 0.21 °Brix and RPD: 6.7 [58]. The calibration models for the determination of pH gave a *R*: 0.85, RMSEC: 0.07 °Brix and RPD: 4.58 for the calibration set and *R*: 0.86, RMSEP : 0.07 °Brix and RPD: 4.4 in validation using samples scanned in reflectance mode [58]. These results highlight the ability of both transmission and reflectance modes to determine the composition of pomegranate fruit samples [58].

The chemical composition of strawberry genotypes using NIR spectroscopy has been determined [59]. The PCA algorithm was applied to interpret spectral differences among the collected samples. Calibration models were developed using PLS regression for the determination of TSS content and firmness [59]. Calibration and validation statistics reported were considered adequate for screening applications [59]. In addition, an on-site system based on NIR spectroscopy was evaluated to predict the maturity of hard green mangoes that are consumed later when ripened off the tree [60]. In this study, a handheld NIR instrument was used to predict DM and starch content in hard green mangoes prior to harvest [60]. The authors of this study concluded that the NIR was able to determine the maturity stages of hard green mangoes on trees [60]. These authors also suggested that NIR spectroscopy can be applied as a decision-making tool to make recommendations about the harvesting data, as well as to inform consumers about the quality of the fruit postharvest [60]. The feasibility of using NIR spectroscopy to predict DM content in 'Hass' avocado at four different times during a growing season has been reported by researchers in New Zealand [61]. The calibration models developed using PLS regression gave a R^2 of 0.88, and a RMSEP of 1.8% DM content [61]. In addition, calibration models based on RMSECV MLR using four wavelengths delivered similar validation statistics as reported for PLS regression [61]. The wavelengths used by the model were associated with C–H absorbance bands (between 820 and 900 nm) [61]. These results suggested that NIR can be used online (e.g. interactance mode) to grade avocado samples on the basis of their DM content to assure taste for the consumers [61]. The prediction of TSS content and pH in cherry fruit samples at different maturities combining NIR wavelengths with hyperspectral imaging has been reported [62]. The hyperspectral image system (874–1734 nm) was used to develop calibration models for TSS content and pH using both PCR and PLS regression [62]. In addition, a genetic algorithm (GA) combined with a successive projections algorithm was applied to identify specific wavelengths before calibration models were developed using MLR regression [62]. Different models were compared where the GA-MLR approach was selected, which delivered an RPD value of 2.7 for TSS content and 2.4 for pH, respectively [62]. The distributions maps for the prediction of both TSS content and pH were created by using the selected NIR bands of each pixel into GA-MLR models [62]. Classification techniques such as LDA were also evaluated where a correct classification ratio of 96.4% was reported by the authors [62].

The determination of vitamin C in intact mandarin fruit samples using VIS-NIR spectroscopy has been reported [63]. The NIR spectra of mandarin collected in reflectance (450 to 1750 nm) were used to develop PLS regression and LSSVM models for the measurement of vitamin C [63]. The calibration models delivered a R^2 0.83, and RMSEP of 2.30 mg/100 g [63].

Data collected using a FT-NIR instrument was compared with the reference data obtained by HPLC for the determination of the concentration of sugars in intact apples [64]. The authors of this study applied different preprocessing techniques such as MSC, the second derivative (Savitsky–Golay), and mean centering normalization, where calibration models were developed using PLS regression [64]. The optimal models delivered a RMSEP and R for glucose, 0.201 and 0.95; fructose, 0.29 and 0.97; sucrose, 0.34 and 0.97, respectively [64]. NIR spectroscopy has been evaluated as tool in breeding programs to improve the efficiency of the breeding by using rapid phenotyping methods (e.g. segregate seedlings for postharvest fruit phenotypes) [65]. A VIS-NIR handheld instrument (Nirvana) was used to collect the spectra of samples before storing apples in cool storage (0.5 °C for a period of 6–10 weeks) and after storage [65]. The NIR spectra were used to develop calibration models for TSS and DM content, as well as fruit firmness. The validation method was achieved by cross-validation, i.e., by leaving out in turn each seedling and then applying the resulting model to predict the postharvest performance of the left-out seedling [65]. The PLS models for the precision of DM content were on average about 2% better than those obtained for the prediction of TSS content, e.g., high genetic correlation (0.90) was observed between these two traits [65]. The RMSEP reported for the determination of TSS and DM content in the seedling samples was 0.80% and 0.70%, respectively, whereas the R^2 was 0.91 and 0.95 for TSS and DM content, respectively [65]. The authors of this study reported an estimated heritability and correlation with sensory properties that were similar to those obtained by the application of NIR spectroscopy [65]. These authors concluded that the models obtained can be used to screen undesirable seedlings, therefore improving the efficiency of the breeding program by not harvesting, cool-storing, and/or undertaking postharvest sensory evaluation of unwanted samples [65].

The use of a FT-NIR (1110–2500 nm) instrument as tool to monitor the vriateal origin of the samples (using ANN), as well as the measurement of sucrose, glucose, and fructose (using PLS regression) in different citrus fruit samples has been evaluated [66]. In this study, ANN combined with GA and correspondences analysis (CA) were used to analyze the spectral data [66]. The PLS regression models developed for sucrose, glucose, and fructose delivered a R^2 ranging between 0.996 and 0.998 [66]. The authors of this study reported excellent classification results of variety and origin using ANN and CA [66].

The feasibility of NIR spectroscopy to evaluate the intra-fruit variability in different apple varieties (e.g. sugar, total polyphenol, and vitamin C content) has been reported [67, 68]. In this study, the authors evaluated different points on the fruit to collect the NIR spectra, e.g., four points 45 from each other in the equatorial region of the apple [67, 68]. The reference values for the quality parameters showed important intra-fruit variability for all quality parameters analyzed where no differences between the mean value at the four points and the mean value of the entire apple [67, 68]. The authors concluded that the mean value obtained from a representative sample of an apple was good enough to assess the variability within the fruit and to estimate, with precision, the content of the quality parameters [67, 68]. The same authors compared the use of a benchtop and portable NIR instruments to evaluate the quality and compositional parameters of apples [67, 68]. In this study, the calibration transfers using a benchtop (XDS) and a handheld spectrometer (MicroNIR) were compared to predict TSS, TA, pulp firmness, and starch-iodine index in apples [67, 68]. The R^2 and RMSECV values obtained using PLS regression were similar between instruments, suggesting that the portable instrument (MicroNIR) offers similar performance to the bench laboratory instrument (XDS) [67, 68]. The statistics obtained in calibration for the prediction of TSS content delivered a $R^2 = 0.89$ with a RMSECV = 0.62 and a $R^2 = 0.91$ with a RMSECV = 0.57 for XDS and MicroNIR, respectively [67, 68]. The authors also transfer the calibration models developed using the spectra of the samples collected in the XDS instrument combined with different regression methods, where the best calibration models reported were obtained using LS-SVM [67, 68]. The Direct Standardization method was used to transfer the calibrations where the RMSEP values obtained after calibration transfer ranged from 74% to 93% using either PLS or MLR regression, respectively. Overall, these results suggest the potential of a simple calibration transfer method to allow the use of historical database registered with benchtop instrument [67, 68].

The utilization of a portable VIS-NIR instrument to predict fruit maturity and anthocyanins in two highbush blueberry (*Vaccinium corymbosum* L.) cultivars (Misty and Sharpblue) has been reported [69]. Correlation for anthocyanins for Misty ($R = 0.97$, $p \leqslant 0.01$) were better than those reported for Sharpblue ($R = 0.71$,

$p>0.05$), while calibrations for TSS ($R = 0.685$, $p \leqslant 0.01$) and fruit firmness ($R = -0.714$, $p \leqslant 0.01$) were only obtained for Sharpblue samples [69].

Miller and Zude [70] investigated the usefulness of NIR spectroscopy to measure TSS content of Florida grapefruit and Honey tangerines. Both online and handheld NIR units were compared. The fruit was subsequently juiced to measure the TSS content of the samples using the refractive index measurement to provide reference data to build the models [70]. Linear regression models were developed between the NIR spectra and TSS content, were the models delivered a $R^2 = 0.67$ in samples from Indian River red grapefruit of small fruit (9.5- to 11.0-cm diameter) tested online [70]. The R^2 values ranged from 0.15 to 0.67. The authors also attempted to classify the samples using a percent correct classification analysis, determining that 9 °Brix was suitable for Interior grapefruit and 10 °Brix was optimal for Indian River red grapefruit [70]. In addition, similar and comparable results were achieved using the online and handheld prototypes, finding slightly higher correlation coefficients for the online equipment [70]. Comparing both systems, higher classification rates in sorting the fruit were obtained for ambient temperature, while decreased correct classification was obtained on prestored fruit (5 °C) [70]. This study assessed the combination of NIR data and information collected by an automated sorting unit for the purpose of classifying fruit based on the TSS content [70]. To achieve this, a neural network (NN) model was developed, which integrated physical and color attributes along with predicted TSS content [70]. Furthermore, an experiment was conducted to classify Indian River red grapefruit over a complete season. The findings of this second year test program revealed that the percent correct classification averaged 85% for TSS content at 10 °sBrix and 79% for an 11 °Brix setpoint [70].

McGlone *et al* [71] reported the effectiveness of VIS-NIR spectroscopy to predict various quality fruit indices such as background color (BC), starch pattern index (SPI), TSS content, penetrometer firmness, quantitative starch, and TA in 'Royal Gala' apples (*Malus domestica* Borkh) collected from eight orchards in two growing regions and two seasons. The results of this study showed that the predictive models were statistically significant in regression terms and accounted for approximately 50%–80% of the data set variance [71]. However, the models developed were not effective in terms of their prediction accuracy when used to analyze new samples [71]. The RPD values obtained were around 3 for all the quality indices predicted, indicating that the models were only suitable for rough screening or grading of the fruit samples [71]. Most of the prediction models showed an influence of the absorbance of chlorophyll, which decreased significantly during the fruit's maturation process, rather than the specific properties of interest [71]. However, the calibration models used to predict TSS content in samples collected after storage were considered to be more reliable by the authors than those collected at harvest time, with an RMSEP of 0.50 and 0.72%, respectively [71]. The authors suggested that this could be associated with the presence of starch in the fruit during harvest, which may have a confounded effect on the TSS calibration models [71].

The use of VIS-NIR spectroscopy has become a prominent non-destructive technique for evaluating fruit quality [72]. The effectiveness of VIS-NIR spectroscopy was used to measure different quality attributes of intact Satsuma mandarin 'Zaojin Jiaogan' (*Citrus reticulata*) and established a correlation between the non-destructive VIS-NIR spectral measurements with different physiological properties of the fruit, including firmness, TSS content, and pH [72]. Calibration models were developed using PCR and PLS regression after preprocessing [72]. The best models for the prediction of TSS content delivered a RMSEP of 0.33 °Brix and a R of 0.94, while for pH and compression force the RMSEP of 0.18 and 8.53, and R values of 0.8 and 0.83, were reported [72].

The use of FT-NIR spectroscopy was also evaluated to predict TSS content and TA in pear samples [73]. Calibration models using the reference data and FT-NIR spectra were achieved using PCR and PLS regression after the use of preprocessing methods (smoothing, MSC, SNV) [73]. The best predictive models delivered an RMSEP of 0.32, 0.02, and R of 0.93, 0.89 for TSS content and TA, respectively [73]. Based on the results, it was concluded that FT-NIR spectrometry could be a fast, reliable, and accurate method for non-invasively predicting the internal quality of pears [73].

Carlini *et al* [74] reported experimental findings regarding the use of PLS regression and wavelength selection for creating models to assess contents content in fruits using VIS-NIR spectroscopy. The authors of this study employed a simple selection method involving subsampling the spectral interval from 600 to 1100 nm with a fixed step to develop accurate models [74]. The SEC and SEP values ranged between 0.50 °Brix and 15 °Brix [74]. However, for the calibration models, the prediction of TSS content in apricot fruit was more challenging and required the selection of individual wavelengths for optimal results, where SEP values at the 0.75 °Brix were reported [74].

Citrus fruit can develop a postharvest disorder called rind physiological (RP), which affects external fruit quality during storage [75]. It typically appears 3–5 weeks after harvest, making it difficult to detect during commercial packing [75]. To address this issue, the use of VIS-NIR spectroscopy and chemometric analysis for non-destructive prediction of RP in 'Marsh' grapefruit has been studied (Citrus × paradisi MacFad) [75]. In this study, VIS-NIR spectroscopy was used to develop calibrations between visual scores of RP and physico-chemical variables related to the disorder and PLS regression. After eight weeks in cold storage and one week in shelf life, the spectral data was correlated with RP scores and rind physico-chemical properties [75]. The study successfully predicted RP (R^2 0.78; RPD: 2.03; RMSEP : 1.41) [75]. In addition, calibration models have been developed to predict total antioxidant capacity (R^2: 0.95), 13 carotene ($R^2 = 0.99$), total carotenoids ($R^2 = 0.92$), chlorophyll a ($R^2 = 0.89$), chlorophyll b ($R^2 = 0.93$), DM ($R^2 = 0.88$), sucrose ($R^2 = 0.91$), glucose ($R^2 = 0.93$), and fructose ($R^2 = 0.94$). In addition, PCA was successful in segregating fruit based on canopy position and susceptibility to RP disorder [75].

In 2022, Li *et al* explored the use of VIS-NIR spectroscopy to characterize the composition and quality of cream strawberry and rabbit-eye blueberry (RB) [76]. Different preprocessing methods were evaluated, i.e., de-trending, moving average smoothing, SNV, and baseline correction, where data dimension was reduced using competitive adaptive reweighted sampling (CARS), successive projection algorithm (SPA), and a combination of both algorithms (CARS+SPA) [76]. The CARS+SPA algorithm delivered the best calibration models R^2 of 0.96, 0.91, 0.91, and 0.93, 0.91, and RPD values of 4.47, 3.28, 3.69, 3.74, and 3.57, respectively. The correlation coefficients between the predicted and measured values of TSS content, TA, and vitamin C content were 0.97, 0.92, 0.97, 0.96, and 0.95, respectively, indicating that the prediction model is highly reliable and stable [76].

7.3.4 Diseases and physiological disorders

In 2015, Lorente *et al* reported the ability of NIR spectroscopy to detect early signs of decay in citrus fruits caused by the *Penicillium digitatum* fungus [77]. The authors recorded the NIR spectra of healthy and decaying tissues from mandarin cv. 'Clemenvilla' using two spectral regions, namely 650 nm to 1050 nm (VIS+NIR) and 1000 nm to 1700 nm (NIR) [77]. Chemometric methods such PCA, FA, and Sammon mapping were used to develop classification models to identify the level of decay in the samples [77]. Successful classification results were obtained using FA based on the NIR spectra, achieving a maximum overall classification accuracy of 97.8% [77]. Overall, the rates of classification achieved were 100% and 94.4% for sound and decaying samples [77]. Teena *et al* reported results from a study in which they used hyperspectral imaging to attempt early detection of microbial infection in dates [78]. The date samples were divided into three groups, namely untreated control (UC), sterile control (SC), and inoculated samples (IS) [78]. The SC group underwent surface sterilization, rinsing, and drying, while the IS group underwent surface sterilization, rinsing, drying, and inoculation with *Aspergillus flavus* [78]. The researchers obtained hyperspectral images of both control samples and the inoculated date fruits by using an area scan hyperspectral imaging system to capture 75 image slices at 10 nm intervals between 960 and 1700 nm every 48 h for 10 days after inoculation [78]. The highest factor loadings of the first principal components (PCs) were used to select the top four most significant wavelengths for feature extraction [78]. Classification techniques such as LDA and quadratic discriminant analysis (QDA) were applied [78]. The classification accuracies of infected samples (IS) were compared separately to uninfected controls (UC) and symptomatic controls (SC) [78]. When comparing IS with SC, the average classification accuracy (using LDA and QDA) was 91.5%, 91.0%, and 99.0%, while the average classification accuracy was 92.4%, 100.0%, and 99.6% when comparing IS with UC [78]. Overall, QDA showed higher accuracy than LDA in all of the tested classification models. However, additional research is necessary to evaluate the effectiveness of this technique on other types of fungal infections and to investigate its impact on the chemical composition of various varieties of date fruit [78].

The feasibility of VIS-NIR spectroscopy was tested to forecast the storage potential of kiwifruit (*Actinidia chinensis* Planch. var. chinensis 'Hort16A') during a 24-week cold storage period (−1.5 °C to 1.5 °C) [79]. To optimize the differentiation between 'sound' fruit and those that developed any disorder, the authors used canonical discriminant analysis (CDA) [79]. Using the CDA classification and sorting process, it was determined that the total occurrence of disorders could have been reduced from 33.9% to 17.9% during the initial harvest and from 14.7% to 8.5% during the second harvest [79]. If the categories were limited to 'sound' fruit and those affected by a single disorder, such as chill injury, then the classification by CDA across all harvests showed that the incidence of disorders could have been reduced from 13.7% to 6.8% [79]. Fruit that developed chill injury and rots were discovered to be less mature at the time of harvest [79]. Based on their NIR profile, the population of affected fruit had lower DM content, significantly lower TSS content, and a greener flesh color than their unaffected counterparts [79]. Based on these results, the authors suggested that in an online setting fruit could be classified into 'sound' and 'affected' groups using NIR spectroscopy, while at harvest NIR spectroscopy can be used to identify the least mature fruit and significantly decrease the occurrence of postharvest storage disorders in the crop [79].

Tomato fruits are frequently affected by fungal diseases, particularly after harvest, which can result in significant losses [80]. Early detection of these diseases may potentially save up to 60% of these losses [80]. The use of VIS-NIR spectroscopy (550–1100 nm) was evaluated to detect gray mold disease at an early stage in tomato fruits [80]. The *Botrytis cinerea* fungus was identified and isolated using two specific primer sets, namely C729+/− and BC108+/BC563−, where three tomato varieties, i.e., Harver, Izmer, and Ekram, were analyzed. Trends in the NIR data were analyzed using PCA, indicating that the VIS-NIR spectra was able to identify infected samples on the second day after infection, even before any visible symptoms appeared on the samples (the first two PCs were able to explain 99% the variance observed) [80].

The use of VIS-NIR spectroscopy was evaluated to detect internal fungal disease caused by *Alternaria alternata* [81]. Reflectance spectra were collected from three areas, i.e., stem-end, equatorial, and stylar-end, of both healthy and infected oranges using a VIS-NIR instrument (400–1100 nm) [81]. The preprocessed wavelengths (second derivative, SNV, mean normalization) were analysed using PCA [81]. The effectiveness of support vector machine and back-propagation neural network (BPNN) classifiers in identifying suitable preprocessors based on the PCs and detecting infected fruits using optimal wavelengths was assessed [81]. The BPNN classifier using optimal wavelengths from the stylar-end area achieved the highest classification accuracy. The classification process included training, validation, and testing stages for the Thompson and Jaffa cultivars. For the Thompson cultivar, the accuracy was 94%, 90%, and 93%, respectively, while for the Jaffa cultivar it was 97%, 91%, and 97%, respectively. These accurate classification results suggested that the models developed were effective in identifying internal fungal infection of *A. alternata* in orange fruit samples. This approach can be extended to other species and varieties [81].

7.3.5 Fruit authenticity, varietal identification, and production system

A procedure has been developed to rapidly determine fruit authenticity and quantification using FT-NIR spectroscopy combined with chemometric techniques

[7, 8]. The procedure is based on analyzing cell wall constituents, such as alcohol-insoluble residue (AIR) and hemicellulose (HC) [82]. The reference values for the content of rhamnose, fucose, arabinose, xylose, mannose, galactose, and glucose in the hemicellulose fraction of apricots, peaches, and pumpkins were obtained through GC [82]. In addition, the spectral data were compared with the fruit content and the gravimetric measurements obtained from sequential fractionation of the AIR [82]. This study also analyzed self-made and commercial apricot and peach fruit preparations, such as jams, and spreads. The NIR data was collected for hemicellulose (n:109) and AIR (n:92) and used to develop calibrations models, between the chemical and gravimetric reference values using PCR regression. The results showed a high level of consistency between the predictions achieved using the FT-NIR method, and were also effective in predicting the neutral sugar composition of the hemicellulose and the fruit content from the AIR [82]. Therefore, the FT-NIR spectroscopy technique proved to be a reliable, fast, and noninvasive method for identifying specific fruits using the spectral data obtained from the hemicellulose fraction and AIR [82]. Hence, FT-NIR showed a great potential as a quality control tool, as well as an alternative technique to GC, which is currently the most commonly used method for analyzing neutral sugars [82]. Additionally, initial research was conducted on the categorization of fruit combinations targeting fruit fraud [82]. In this study, fruit samples that were considered fraudulently mixed with authentic fruits, such as peach or pumpkin into apricot, or pumpkin into peach, the likelihood of them being classified as a specific fruit was considerably reduced or eliminated [82].

China cultivates over 1000 varieties of peach, which exhibit significant variations in quality (e.g. sugars and DM) [83]. Accurate differentiation of peach varieties is essential not only for peach traders but also along the entire supply and value chain to assure the quality and safety of these fruits to the consumer [83]. The identification of the identity (variety) and origin of the fruit has been among the most important issues facing the fruit and food industries in recent years [7, 8, 83]. The development of a detection model requires the users to have sufficient expertize and knowledge to develop complex analytical systems [83]. The differentiation of peach varieties has been evaluated using VIS-NIR spectroscopy (350–820 nm), where these researchers proposed a deep learning-based detection approach for the classification of peach varieties through the analysis of the VIS-NIR spectra [83]. This method utilizes a one-dimensional convolutional NN and a VIS-NIR spectra database consisting of five different types of peaches to achieve multi-identification of peach varieties [83]. The network model is trained and subsequently applied to predict the test data set [83]. The accuracy of the deep learning-based models achieved 100% in the validation data set and 94.4% in the test data set. Overall, this study highlights the successful differentiation of peach varieties using VIS-NIR spectroscopy and deep learning [83].

The ability of NIR spectroscopy to differentiate between organic and inorganic pineapple fruits, and to predict TSS content has also been reported [84]. In this study, a portable NIR spectrometer combined with PCA, KNN, and LDA, as well as different preprocessing methods were evaluated to differentiate between classes (organic and conventional) of pineapple fruit samples. Meanwhile, PLS regression was used to predict the TSS content of the samples [84]. A model developed by combining MSC-PCA-LDA accurately identified 100% of the organic samples from conventionally fruit samples. The prediction of TSS content was achieved by a combination of MSC with PLS regression, $R^2 = 0.851$ and RMSEC = 0.950 °Brix, and $R^2 = 0.854$ and RMSEP = 0.842 °Brix, prediction and calibration, respectively [84].

7.4 Final considerations

The applications and examples reported above show the feasibility of NIR spectroscopy to measure and monitor the different steps in the supply and value chain of fruits. Most of these reports considered NIR spectroscopy as a quick and user-friendly method, and it has already achieved significant industrial recognition during the routine analysis of fruits such as mango, avocado, apples, and citrus. However, the widespread adoption of NIR spectroscopy in the industry may be impeded by several critical factors and limitations, including the availability and cost of instruments, the type of application to be implemented, and the level of understanding of the technology. At present, several companies are manufacturing NIR instruments that provide platforms such as benchtop, portable, as well as online instruments. Defining the key specifications for a given instrument, such as the wavelength or wavenumber scanning range (e.g. VIS plus NIR, NIR), the number of data points, the signal-to-noise ratio level required, the instrument stability and measurement time (e.g. temperature), as well as the type of sample or application, is essential before selecting a suitable NIR instrument to be purchased. In general terms, the selection of the appropriate instrument type is primarily determined by the sample (e.g. whole fruit, juice, homogenated sample) and the parameter(s) to be measured, which heavily depends on the nature and chemical concentration of the compounds to be determined or measured in the sample.

One of the most important aspects to consider before and during the application and implementation of NIR spectroscopy is the need for proper training. For instance, while routine analyzes can be conducted by an analysts with high school education or training, both the development of a calibration and the implementation of the method require different skills and knowledge (e.g. chemometrics, optics, spectroscopy). As stated in the previous sections, developing a calibration is one of the critical aspects required in NIR spectroscopy implementation, which demands high levels of expertize, especially in chemometrics or multivariate data analysis, to ensure the success of an application.

While NIR spectroscopy may not be able to measure molecules with low concentration, the overall effects or collinearities in the fruit matrix can be observed or evaluated through the NIR spectrum of a specific sample. One of the primary benefits of using the NIR spectroscopy combined with chemometric techniques, such as PCA or discriminant analysis, is that it can help uncover specific compositional traits that are not readily identifiable through conventional targeted chemical analysis.

The considerable cost and time savings, coupled with the environmentally-friendly nature of the technology, have made NIR spectroscopy a highly desirable technique to be used along the supply and value fruit chain. The extensive range of applications of NIR spectroscopy, whether in routine practice or development, continues to expand. Nonetheless, the absence of formal education in both NIR spectroscopy (e.g. spectra interpretation, sample presentation) and chemometrics (e.g. calibration development) remains a hurdle for the widespread adoption of this technology.

References

[1] Pennington J A T and Fisher R A 2009 Classification of fruits and vegetables *J. Food Compos. Anal.* **22S** S23–31

[2] W H O 2016 Promoting fruit and vegetable consumption around the world (http://who.int/dietphysicalactivity/fruit/en/) (accessed 30 March 2016).

[3] Dembitsky V M, Poovarodom S, Leontowicz H, Leontowicz M, Vearasilp S and Trakhtenberg S *et al* 2011 The multiple nutrition properties of some exotic fruits: Biological activity and active metabolites *Food Res. Int.* **44** 1671–701

[4] Moure A, Cruz J M, Franco D, Domínguez J M, Sineiro J and Domínguez H *et al* 2001 Natural antioxidants from residual sources *Food Chem.* **72** 145–17

[5] Schieber A, Stintzing F and Carle R 2001 By-products of plant food processing as a source of functional compounds—recent developments *Trends Food Sci. Technol.* **12** 401–13

[6] Cozzolino D 2012 Recent trends on the use of infrared spectroscopy to trace and authenticate natural and agricultural food products *Appl. Spectrosc. Rev.* **47** 518–30

[7] Arvantoyannis I, Katsota M N, Psarra P, Soufleros E and Kallinthraka S 1999 Application of quality control methods for assessing wine authenticity: use of multivariate analysis (chemometrics) *Trends Food Sci. Technol.* **10** 321–6

[8] Woodcock T, Downey G and O'Donnell C P 2008 Better quality food and beverages: the role of near infrared spectroscopy *J. Near Infrared Spectros.* **16** 1–29

[9] Krüger H and Schulz H 2007 Analytical techniques for medicinal and aromatic plants *Stewart Post Harvest Rev.* **3** 1–12

[10] Schulz H 2004 Analysis of coffee, tea, cocoa, tobacco, spices, medicinal and aromatic plants, and related products In C A Roberts, J Workman and J B Reeves III *Near Infrared Spectroscopy in Agriculture* (Madison: ASA, CSSA, SSSA) 345–57

[11] McGoverin C M, Weeranantanaphan J, Downey G and Manley M 2010 The application of near infrared spectroscopy to the measurement of bioactive compounds in food commodities *J. Near Infrared Spectrosc.* **18** 87–111

[12] Cozzolino D, Shah N, Cynkar W and Smith P 2011 Technical solutions for analysis of grape juice, must and wine: the role of infrared spectroscopy and chemometrics *Anal. Bioanal. Chem.* **401** 1479–88

[13] Van Duyn M A S and Pivonka E 2000 Overview of the health benefits of fruit and vegetable consumption for the dietetics professional *J. Am. Diet Assoc.* **100** 1511–21

[14] Wootton-Beard P C and Ryan L 2011 Improving public health? The role of antioxidant-rich fruit and vegetable beverages *Food Res. Int.* **44** 3135–48

[15] Delgado-Vargas F, Jimenez A R and Paredes-Lopez O 2000 Natural pigments: carotenoids, anthocyanins, and betalains-characteristics, biosynthesis, processing, and stability *Crit. Rev. Food Sci. Nutr.* **40** 173–289

[16] Choe E and Min D B 2006 Chemistry and reactions of reactive oxygen species in foods *Crit. Rev. Food Sci. Nutr.* **46** 1–22
[17] Wang S, Melnyk J P, Tsao R and Marcone M F 2011 How natural dietary antioxidants in fruits, vegetables and legumes promote vascular health *Food Res. Int.* **44** 14–22
[18] Aberoumand A 2011 A review article on edible pigments properties and sources as natural biocolorants in foodstuff and food industry *World J. Dairy Food Sci.* **6** 71–8
[19] Wrolstad R E and Culver C A 2012 Alternatives to those artificial FD&C food colorants *Annu. Rev. Food Sci. Technol.* **2** 59–77
[20] Robert L, Rickey Y, Marvin A and Alex S 1987 Anthocyanins as food colorants: a review *J. Food Biochem.* **11** 201–47
[21] Wissgott U and Bortlik K 1996 Prospects for new natural food colorants *Trends Food Sci. Technol.* **7** 298–302
[22] Sumner L W, Mendes P and Dixon R A 2003 Plant metabolomics: large-scale phytochemistry in the functional genomics era *Phytochemistry* **62** 817–36
[23] Cozzolino D, Cynkar W, Janik L, Dambergs R G and Gishen M 2006 Analysis of grape and wine by near infrared spectroscopy—a review *J. Near Infrared Spectrosc.* **14** 279–89
[24] Cozzolino D 2009 Near infrared spectroscopy in natural products analysis *Planta Med.* **75** 746–57
[25] Martson A and Hostettmann K 2009 Natural product analyses over the last decades *Planta Med.* **75** 672–83
[26] Karoui R, Downey G and Blecker C 2010 Mid-infrared spectroscopy coupled with chemometrics: a tool for the analysis of intact food systems and the exploration of their molecular structure-quality relationships—a review *Chem. Rev.* **110** 6144–68
[27] Subramanian A and Rodrigez-Saona L 2009 Fourier transform infrared (FTIR) spectroscopy In *Infrared Spectroscopy for Food Quality Analysis and Control* ed Da Wen Sun (Oxford: Academic) pp 146–74
[28] Li-Chan E C Y 2010 Introduction to vibrational spectroscopy In *Applications of Vibrational Spectroscopy in Food Science* ed E Li-Chan, P R Griffiths and J M Chalmers (Wiley and Sons)
[29] Schulz H, Baranska M and Baranski R 2005 Potential of NIR-FT-Raman spectroscopy in natural carotenoid analysis *Biopolymers* **77** 212–21
[30] Burns D A and Margoshes M 1992 Historical development In: *Handbook of Near Infrared Analysis* ed D A Burns and E W Ciurczak (New York: Marcel Dekker) pp 1–11
[31] Butler L A 1983 The history and background of NIR *Cereal Food World* **3** 238–40
[32] Baranska M and Schultz H 2006 Application of infrared and Raman spectroscopy for analysis of selected medicinal and spice plants *J. Med. Spice Plant (Z. Arzn. Gew. Pfl.)* **2** 72–80
[33] Baranska M, Schulz H, Rösch P, Strehle M A and Popp J 2004 Identification of secondary metabolites in medicinal and spice plants by NIR-FT-Raman microspectroscopic mapping *Analyst* **129** 926–30
[34] Crocombe R A 2004 MEMS technology moves process spectroscopy into a new dimension *Spectrosc Europe* **3** 16–9
[35] Sorak D, Herberholz L, Iwascek S, Altinpinar S, Pfeifer F and Siesler H W 2012 New developments and applications of handheld Raman, mid-infrared, and near-infrared spectrometers *Appl. Spectrosc. Rev.* **47** 83–115

[36] Gottlieb D M, Schultz J, Bruun S W, Jacobsen S and Søndergaard I b 2004 Multivariate approaches in plant science *Phytochemistry* **65** 1531–48
[37] Wold S 1995 Chemometrics; what do we mean with it, and what do we want from it? *Chem. Intell. Lab. Syst.* **30** 109–15
[38] Naes T, Isaksson T, Fearn T and Davies T 2002 *A User-Friendly Guide to Multivariate Calibration and Classification* (Chichester: NIR Publications)
[39] Otto M 1999 *Chemometrics: Statistics and Computer Application in Analytical Chemistry* (Chichester: Wiley-VCH)
[40] Brereton R G 2003 *Chemometrics: Data Analysis for the Laboratory and Chemical Plant* (Chichester: Wiley)
[41] Brereton R G 2007 *Applied Chemometrics for Scientist* (Chichester: Wiley)
[42] Cozzolino D, Shah N, Cynkar W and Smith P 2011 A practical overview of multivariate data analysis applied to spectroscopy *Food Res. Int.* **44** 1888–96
[43] Roggo Y, Chalus P, Maurer L, Lema-Martinez C, Edmond A and Jent N 2007 A review of near infrared spectroscopy and chemometrics in pharmaceutical technologies *J. Pharm. Biomed. Anal.* **44** 683–90
[44] Huang H, Yu H, Xu H and Ying Y 2008 Near infrared spectroscopy for on/in-line monitoring of quality in foods and beverages: a review *J. Food Eng.* **87** 303–13
[45] Stratis D N, Eland K L, Carter J C, Tomlinson S J and Angel S M 2001 Comparison of acoustic-optic and liquid crystal tunable filters for laser induced breakdown spectroscopy *Appl. Spectrosc.* **55** 999–1004
[46] Williams P C 2001 Implementation of near-infrared technology, In *Near Infrared Technology in the Agricultural and Food Industries* ed P C Williams and K H Norris (St. Paul, MN: American Association of Cereal Chemist) pp 145–69
[47] Esbensen K H 2002 *Multivariate Data Analysis in Practice* (Oslo, Norway: CAMO Process AS)
[48] Norris K H and Ritchie G E 2008 Assuring specificity for a multivariate near-infrared (NIR) calibration: the example of the Chambersburg Shoot-out 2002 data set *J. Pharmacol. Biomed. Anal.* **48** 1037–41
[49] Fearn T 2002 Assessing calibrations: SEP, RPD, RER and R2 *NIR News* **13** 12–4
[50] Nicolai B M, Beullens K, Bobelyn E, Peirs A, Saeys W, Theron K I and Lammertyn J 2007 Nondestructive measurement of fruit and vegetable quality by means of NIR spectroscopy: a review *Postharvest Biol. Technol.* **46** 99–118
[51] Wang H, Peng J, Xie C, Bao Y and He Y 2015 Fruit quality evaluation using spectroscopy technology: a review *Sensors* **15** 11889–927
[52] Shah S S, Zeb A, Qureshi W S, Arslan M, Malik A U, Alasmary W and Alanazi E 2020 Towards fruit maturity estimation using NIR spectroscopy *Infrared Phys. Technol.* **111** 103479
[53] Liu Y, Sun X and Ouyang A 2010 Nondestructive measurement of soluble solid content of navel orange fruit by visible–NIR spectrometric technique with PLSR and PCA-BPNN *LWT-Food Sci. Technol.* **43** 602–7
[54] Rodriguez-Saona L E, Fry F S, McLaughlin M A and Calvey E M 2001 Rapid analysis of sugars in fruit juices by FT-NIR spectroscopy *Carbohydr. Res.* **336** 63–74
[55] Alamar P D, Caramês E T, Poppi R J and Pallone J A 2016 Quality evaluation of frozen guava and yellow passion fruit pulps by NIR spectroscopy and chemometrics *Food Res. Int.* **85** 209–14

[56] McGlone V A, Fraser D G, Jordan R B and Künnemeyer R 2003 Internal quality assessment of mandarin fruit by vis/NIR spectroscopy *J. Near Infrared Spectrosc.* **11** 323–32
[57] Cayuela J A and Weiland C 2010 Intact orange quality prediction with two portable NIR spectrometers *Postharvest Biol. Technol.* **58** 113–20
[58] Khodabakhshian R, Emadi B, Khojastehpour M and Golzarian M r 2019 A comparative study of reflectance and transmittance modes of Vis/NIR spectroscopy used in determining internal quality attributes in pomegranate fruits *J. Food Meas. Charact.* **13** 3130–9
[59] Mancini M, Mazzoni L, Gagliardi F, Balducci F, Duca D, Toscano G, Mezzetti B and Capocasa F 2020 Application of the non-destructive NIR technique for the evaluation of strawberry fruits quality parameters *Foods.* **9** 441
[60] Saranwong I, Sornsrivichai J and Kawano S 2003 On-tree evaluation of harvesting quality of mango fruit using a hand-held NIR instrument *J. Near Infrared Spectrosc.* **11** 283–93
[61] Clark C J, McGlone V A, Requejo C, White A and Woolf A B 2003 Dry matter determination in 'Hass' avocado by NIR spectroscopy *Postharvest Biol. Technol.* **29** 301–8
[62] Li X, Wei Y, Xu J, Feng X, Wu F, Zhou R, Jin J, Xu K, Yu X and He Y 2018 SSC and pH for sweet assessment and maturity classification of harvested cherry fruit based on NIR hyperspectral imaging technology *Postharvest Biol. Technol.* **143** 112–8
[63] Sun X, Dong X, Cai L, Hao Y, Ouyang A and Liu Y 2012 Visible-NIR spectroscopy and least square support vector machines regression for determination of vitamin C of mandarin fruit *Sens. Lett.* **10** 506–10
[64] Liu Y, Ying Y, Yu H and Fu X 2006 Comparison of the HPLC method and FT-NIR analysis for quantification of glucose, fructose, and sucrose in intact apple fruits *J. Agric. Food Chem.* **54** 2810–5
[65] Kumar S, McGlone A, Whitworth C and Volz R 2015 Postharvest performance of apple phenotypes predicted by near-infrared (NIR) spectral analysis *Postharvest Biol. Technol.* **100** 16–22
[66] Tewari J C, Dixit V, Cho B K and Malik K A 2008 Determination of origin and sugars of citrus fruits using genetic algorithm, correspondence analysis and partial least square combined with fiber optic NIR spectroscopy *Spectrochim. Acta, Part* A **71** 1119–27
[67] Pissard A, Baeten V, Romnée J M, Dupont P, Mouteau A and Lateur M 2012 Classical and NIR measurements of the quality and nutritional parameters of apples: a methodological study of intra-fruit variability *BASE* **16** pp 294–306
[68] Pissard A, Marques E J, Dardenne P, Lateur M, Pasquini C, Pimentel M F, Pierna J A and Baeten V 2021 Evaluation of a handheld ultra-compact NIR spectrometer for rapid and non-destructive determination of apple fruit quality *Postharvest Biol. Technol.* **172** 111375
[69] Ribera-Fonseca A, Noferini M and Rombolá A D 2016 Non destructive assessment of highbush blueberry fruit maturity parameters and anthocyanins by using a visible/near infrared (vis/NIR) spectroscopy device: a preliminary approach *J. Soil Sci. Plant Nutr.* **16** 174–86
[70] Miller W M and Zude-Sasse M 2004 NIR-based sensing to measure soluble solids content of *Florida citrus Appl. Eng. Agric.* **20** 321
[71] McGlone V A, Jordan R B and Martinsen P J 2002 Vis/NIR estimation at harvest of pre-and post-storage quality indices for 'Royal Gala'apple *Postharvest Biol. Technol.* **25** 135–44
[72] Gómez A H, He Y and Pereira A G 2006 Non-destructive measurement of acidity, soluble solids and firmness of Satsuma mandarin using Vis/NIR-spectroscopy techniques *J. Food Eng.* **77** 313–9

[73] Liu Y and Ying Y 2007 Noninvasive method for internal quality evaluation of pear fruit using fiber-optic FT-NIR spectrometry *Int. J. Food Prop.* **10** 877–86

[74] Carlini P, Massantini R and Mencarelli F 2000 Vis-NIR measurement of soluble solids in cherry and apricot by PLS regression and wavelength selection *J. Agric. Food Chem.* **48** 5236–42

[75] Ncama K, Tesfay S Z, Fawole O A, Opara U L and Magwaza L S 2018 Non-destructive prediction of 'Marsh'grapefruit susceptibility to postharvest rind pitting disorder using reflectance Vis/NIR spectroscopy *Sci. Hortic.* **231** 265–71

[76] Li L, Hu D Y, Tang T Y and Tang Y L 2023 Non-destructive detection of the quality attributes of fruits by visible-near infrared spectroscopy *J. Food Meas. Char.* 1526–34

[77] Lorente D, Escandell-Montero P, Cubero S, Gómez-Sanchís J and Blasco J 2015 Visible–NIR reflectance spectroscopy and manifold learning methods applied to the detection of fungal infections on citrus fruit *J. Food Eng.* **163** 17–24

[78] Teena M A, Manickavasagan A, Ravikanth L and Jayas 2014 DS. Near infrared (NIR) hyperspectral imaging to classify fungal infected date fruits *J. Stored Prod. Res.* **59** 306–13

[79] Clark C J, McGlone V A, De Silva H N, Manning M A, Burdon J and Mowat A D 2004 Prediction of storage disorders of kiwifruit (*Actinidia chinensis*) based on visible-NIR spectral characteristics at harvest *Postharvest Biol. Technol.* **32** 147–58

[80] Najjar K and Abu-Khalaf N 2021 Visible/near-infrared (VIS/NIR) spectroscopy technique to detect gray mold disease in the early stages of tomato fruit: VIS/NIR spectroscopy for detecting gray mold in tomato *J. Microbiol., Biotechnol. Food Sci.* **11** e3108

[81] Ghooshkhaneh N G, Golzarian M R and Mollazade K 2023 VIS-NIR spectroscopy for detection of citrus core rot caused by *Alternaria alternata Food Control* **144** 109320

[82] Kurz C, Leitenberger M, Carle R and Schieber A 2010 Evaluation of fruit authenticity and determination of the fruit content of fruit products using FT-NIR spectroscopy of cell wall components *Food Chem.* **119** 806–12

[83] Rong D, Wang H, Ying Y, Zhang Z and Zhang Y 2020 Peach variety detection using VIS-NIR spectroscopy and deep learning *Comput. Electron. Agric.* **175** 105553

[84] Amuah C L, Teye E, Lamptey F P, Nyandey K, Opoku-Ansah J and Adueming P O 2019 Feasibility study of the use of handheld NIR spectrometer for simultaneous authentication and quantification of quality parameters in intact pineapple fruits *J. Spectrosc.* **5975461**

IOP Publishing

Advances in Spectroscopic Analysis of Food and Drink

Ashutosh Kumar Shukla

Chapter 8

Combining compositional characteristics, smartphone image, and NIR spectroscopy in a data fusion approach to study relationships between traditional infusions

Michel Rocha Baqueta, Grasieli Beloni de Melo, Elem Tamirys dos Santos Caramês, Patrícia Valderrama and Juliana Azevedo Lima Pallone

Coffee, mate tea, and star anise are widely used in the preparation of infusions and are clearly distinguishable. The objective of this chapter is to evaluate relationships between powders of coffees (espresso, traditional, and instant), mate tea, star anise, and their respective infusions from the combination of the nutritional composition, bioactive potential, smartphone image, and near-infrared (NIR) spectroscopy in a chemometric data fusion approach. The results show that the proposed method is useful to find the relationship between traditional infusions through NIR vibrations, nutritional composition, bioactive potential, and smartphone image evaluated in a combined manner with principal component analysis (PCA).

8.1 Introduction

Simultaneous analysis is increasingly promoting the development of studies to understand the interesting relationships between complex matrices. For these reasons, multiplatform strategies have proven to be a great solution for the simultaneous evaluation of distinct foods.

Coffee and its variations, including traditional, instant, and espresso (capsules), are consumed globally thanks to their unique flavor and aroma, as well as sense of emotional well-being. In the past few years, more than two billion cups of coffee have been consumed worldwide every day [1]. Nevertheless, other alcohol-free preparations are also consumed as infusions, here generalized as 'brew'. For

doi:10.1088/978-0-7503-5573-5ch8 © IOP Publishing Ltd 2023

example, mate tea is widely consumed in South America, which is prepared from the dried leaves of *Ilex paraguariensis* [2, 3]. Another known brew consumed in Asia and Northern America is prepared from the fruits of *Illicium verum* Hook. f., a traditional aromatic spice popularly referred to as star anise [4]. The widespread consumption of these traditional infusions, either for cultural reasons or nutraceutical benefits, has motivated scientists to find out more about these infusions, and their nutritional and bioactive potential for consumers.

Coffee, mate tea, and star anise are known for their stimulant, beneficial, and nutritional properties. They are considered to be rich sources of bioactive and nutritional compounds, and several studies have associated these compounds with health benefits [3–7]. However, their leachability rates may vary in the brews, so it is necessary to understand them in more depth.

Among the available technologies for the analysis of alcohol-free beverages, vibrational-infrared spectroscopy, atomic spectrometry, and computer vision systems using smartphone-based images have many potential applications. Except for atomic spectrometry, which is more useful in elemental analysis [8, 9], these techniques have been reported in the literature as alternatives to traditional methods due to their simplicity, speed, and eco-friendliness [10–13].

Currently, there are several approaches for combining results analyzed simultaneously in a unique sample set, e.g., coffees, mate tea, star anise, and their brews. In particular, the application of data fusion strategies has significantly increased and shown good results in different areas. There are three data fusion levels (low, mid, and high), more details about them can be found in the literature [14–16]. In this study, low-level fusion was implemented and briefly consists of concatenating the data from different analytical sources after some preprocessing steps into a common data matrix after suitable preprocessing and weighting [14].

To process and interpret this multiple information, and to find interesting relationships between samples, chemometric tools are necessary to extract the highest possible information content. PCA is a well-established unsupervised pattern recognition tool. It is known to reduce data dimensionality, it is useful in exploratory data analysis, and it can be used to identify patterns in complex data. Combined with data fusion, PCA has been useful to evaluate biochemical changes during the storage of extra virgin olive oil [17] and to monitor the kombucha fermentation process [18].

The objective of this chapter is to evaluate and study the relationships between powders of coffees (espresso, traditional and instant), mate tea, star anise, and their respective infusions from a combination of the nutritional composition, bioactive potential, smartphone image, and near-infrared (NIR) spectroscopy in a data fusion approach. The results were individually analyzed and then combined by a low-level data fusion coupled with subsequent chemometric interpretation via PCA. It is important to highlight that a data fusion approach is used for the first time to study the relationships between coffee, mate tea, and star anise preparations.

8.2 Materials and methods

8.2.1 Reagents and samples

Nitric acid (⩾65%, Sigma-Aldrich, St. Louis, USA) and standard solutions (1000 mg l^{-1}) of calcium (Ca), magnesium (Mg), iron (Fe), zinc (Zn), potassium (K), copper (Cu), and manganese (Mn), as well as Folin–Ciocalteu's phenol reagent (2 M) and gallic acid (⩾98%) were purchased from Sigma-Aldrich (St. Louis, USA). Hydrogen peroxide (29%) and sodium carbonate (Na_2CO_3, 99.8%) were purchased from Synth (São Paulo, Brazil). De-ionized water (ultrapure) was supplied by a Sartorius–Arium Comfort water purification system and used in all of the experiments, including brewing preparations.

The samples were purchased in the markets at Campinas, São Paulo, Brazil and included star anise (*Illicium verum* Hook. f.) (sample 1), mate tea (*Ilex paraguariensis*) (sample 2), instant coffee (sample 3), traditional roasted and ground coffee (sample 4), and two espresso coffee capsules from different producers (samples 5 and 6). Coffee samples were blends of *Coffea arabica*/*C. canephora* var. robusta and the concentration of each variety was not informed. To eliminate or reduce possible particle size effects in the experiments, the samples were ground (Biovera, Brazil, model Ika A11) and sieved through an 80-mesh sieve (180 μm).

8.2.2 Brewing preparation

The brews were prepared according to the sample's label instructions using de-ionized water to reduce water effects in the mineral extraction. Briefly, star anise and mate tea (about 2 g) were placed in 400-mL glass beakers, poured with 200 ml of de-ionized water (95°C) at room temperature (22 ± 1°C) and left under cover to brew for 10 min. The same process was repeated to intact star anise to check the effect of not grinding star anise fruits in the extraction of minerals and phenolic compounds. In the coffees (instant and traditional), the brews were prepared using about 1 g and 8.5 g to 50 and 100 ml of de-ionized water (95°C), respectively. For espresso coffees, commercial capsules (about 5 or 7.5 g of roasted and ground coffee in each) were prepared using coffee machines with about 40 ml of de-ionized water. The temperature of the espressos at the time of preparation was about 75°C. All brews were filtered through a quantitative filter paper (Nalgon, 9 cm diameter).

8.2.3 Extraction and evaluation of total phenolic compounds

Phenolic compounds were extracted from each solid sample according to the method proposed in previous studies [19] with modifications. About 0.05 g of solid samples were extracted with 8 ml of water:acetone (50:50 v/v) in a household microwave oven (CCE, Brazil, model MW-1580) operating at 1420 W. The extraction was made in a vortex for 10 s and two cycles of 30 s of irradiation with a cooling time of 10 min (until 25°C) between cycles to avoid thermal degradation. The samples were then centrifuged at 6000 rpm for 10 min at 4°C (Solab Cientifica, Brazil, model SL-706). The supernatant was collected and used in the total phenolic content (TPC) assay. For brews, a direct analysis was performed because the goal was to evaluate their total phenolic levels.

TPC in solid and brew samples was determined by the Folin–Ciocalteu method [20]. In a dark room, 85 μl of each extracted sample with 43 μl of Folin–Ciocalteau reagent (1 N converter para mol l^{-1}) and 212 μl of Na_2CO_3 (75 g l^{-1}) was then added to each sample in a 96-well microplate. The absorbance was measured at 725 nm after 25 min of incubation at room temperature in a multi-mode microplate reader (BMG Labtech, Germany, model Fluostar Omega). The TPC concentrations were obtained from a standard curve of gallic acid (1–50 mg ml^{-1}) and the results were expressed as mg of gallic acid equivalents (GAE) per 1000 ml of sample. The analyzes were performed in triplicate.

8.2.4 Determining the essential minerals

The mineralization of the solid samples and their brews (0.6 g of powders and 2 ml of brews) was performed using a digester block (M242 model, Quimis, Brazil). Nitric acid and hydrogen peroxide in total proportions of 6 ml and 2 ml, respectively, were used. Mineralization time was performed during a total of four hours in two cycles: 2 h at 110°C and 2 h at 130°C. Then, mineralized samples were filtered through quantitative filter paper (Nalgon, 9 cm diameter) and diluted in relation to the concentration of each element. The essential minerals (Ca, Mg, Fe, Zn, K, Cu, and Mn) were determined using a flame atomic absorption spectrometer (FAAS), AAnalyst 200 model, with a deuterium lamp for correction of the background radiation and hollow cathode lamps (Perkin Elmer). The analyzes were performed in triplicate.

8.2.5 NIR spectra acquisition

NIR spectra data were recorded in the solid and brew samples. For the acquisition of NIR spectra from solid samples, about 1 g of each sample was placed into a glass cuvette and acquisition performed on reflectance mode. For brews, the samples were placed into a petri dish, and a liquid reflector (Perkin Elmer, Walthman, USA, part number L118–0503) was placed above the sample and slowly placed against the glass before scanning. NIR spectra acquisition from liquid samples was performed in transfectance mode. The near-infrared reflectance accessory (Perkin Elmer, Walthman, USA, model NIRA) was coupled in the FT-NIR spectrometer (Perkin Elmer, Walthman, USA, model Spectrum 100 N) using indium gallium arsenide (InGaAs) as a detector, in the range from 4000 to 10 000 cm^{-1}, with the mean spectra of 32 scans and a resolution of 4 cm^{-1}. Both spectra were recorded in absorbance. The analyzes were performed in triplicate.

8.2.6 Smartphone-based image

Solid and brews samples were evaluated by a smartphone (Samsung Galaxy S5 SM-G9000, 16-megapixel camera, resolution of 5312×2988 pixel, autofocus, equipped with Android 6.0.1.) with a free accessibility application (app) named Color Grab (version 3.6.1, Loomatix©). Capture image area, illumination, and focal distance were kept constant in all experiments. After the image capture, the values of the individual components (red, green, and blue; RGB) give the mean values for all pixels. Thus, the reflected radiation was measured, yielding values of 0 (total absorption of incident radiation) to 255 (negligible radiation absorption and maximum reflection). The reflectance values measured in these channels can then be taken as the analytic signal. For the data acquisition, about 1 g (solid) or 2 ml (brew) of each sample was placed into a petri dish and positioned in the central region of the Color Grab scanner app. Measurements were acquired with a smartphone's rear camera with a focal length of 18.0 cm from the camera to the petri dish. The measurements were obtained in a room with artificial lights (lamps) in triplicate.

8.2.7 Data fusion and PCA

Initially, all of the variables studied were organized in matrices where the lines represented the samples and the columns represented the variables. In this study, each replicate of analysis was considered to be one sample, resulting in a total of 36 samples. Samples were numbered from 1 to 36, comprising solid (1–18) and their brews (19–36). The samples were organized in the following ranking: star anise, mate tea, instant coffee, espresso coffee 1, espresso coffee 2, and traditional coffee. Low-level data fusion was applied to a matrix (36 samples per 6012 variables) composed of all of the studied variables, including essential minerals (seven variables, i.e., Ca, Mg, Fe, Zn, K, Cu, and Mn), total phenolic content (one variable), NIR spectroscopy (6001 variables), and RGB mean values (three variables).

In the development of low-level data fusion, each data set was previously preprocessed before combining the fusion matrix. Essential minerals, total phenolic content, and RGB were autoscaled, while NIR spectra were pretreated by multiplicative scatter correction [21] and Savitzky–Golay smoothing with a window size of 5 points and first-order polynomial [22]. After individual preprocessing, both data sets were individually normalized and joined in a fusion matrix with all variables. Since the results from analyzes were measured using very different entities or units, the data sets were normalized as in previous studies because the measured features show a difference in variances [17]. After data merging, the fusion matrix was evaluated by the PCA with mean center preprocess. PCA was used to explore the data structure, and to trace patterns of similarity between solid and brew samples.

8.3 Results and discussions

8.3.1 Bioactive and nutritional compounds

The total mineral concentrations (Mg, Ca, K, Fe, Mn, Zn, and Cu) in the solid and brews are shown in table 8.1. Mg, Ca, and K were the most abundant elements in all samples and presented an average concentration of 9103.84, 1645.80, and 1021.18 mg kg^{-1}, respectively. Fe and Mn presented moderate concentrations (average of 305.52 and 85.86 mg kg^{-1}, respectively), and Zn (19.53 mg kg^{-1}) and Cu (10.99 mg kg^{-1}) showed the lowest concentrations. Even though each sample has a mineral

Table 8.1. Concentrations of essential minerals (mg/kg) and TPC (mg GAEq/L) in the solid samples and their brews.

Sample	Type	Mg	Ca	K	Fe	Mn	Zn	Cu	Sum[a]	TPC
Star anise	Solid	741.53 ± 17.29	1183.55 ± 12.86	8484.86 ± 372.54	131.31 ± 3.86	2414.31 ± 21.81	17.66 ± 1.29	7.30 ± 0.30	12 980.51	43.09 ± 1.66
	Brew	10.14 ± 1.59	9.31 ± 0.62	149.44 ± 0.39	<2.28	3.74 ± 0.26	<3.74	<0.47	172.63	26.37 ± 1.37
	Not milled	1.74 ± 0.12	86.40 ± 6.70	56.92 ± 2.22	<2.28	<0.53	<3.74	<0.47	145.06	11.15 ± 0.40
Mate tea	Solid	10 394.07 ± 508.76	7491.69 ± 429.31	13 596.10 ± 841.26	400.39 ± 18.14	217.42 ± 14.93	69.95 ± 0.88	11.56 ± 0.18	32 181.18	272.85 ± 1.08
	Brew	104.21 ± 0.96	31.04 ± 2.89	339.22 ± 9.09	10.92 ± 21.18	1.58 ± 0.46	<3.74	<0.47	486.97	108.07 ± 2.88
Espresso coffee 1	Solid	2029.66 ± 106.44	989.32 ± 13.75	25 026.29 ± 288.84	30.01 ± 1.64	67.96 ± 1.73	3.97 ± 0.20	15.24 ± 0.27	28 162.45	923.85 ± 39.15
	Brew	109.19 ± 1.70	30.13 ± 0.66	1964.71 ± 36.04	<2.28	<0.53	<3.74	<0.47	2104.03	61.36 ± 1.19
Espresso coffee 2	Solid	1888.65 ± 97.66	1074.61 ± 29.53	21 873.64 ± 773.59	29.67 ± 1.37	56.47 ± 2.52	3.97 ± 0.42	14.88 ± 0.08	24 941.89	927.30 ± 39.15
	Brew	97.66 ± 1.70	49.94 ± 1.46	1778.65 ± 60.70	<2.28	<0.53	<3.74	<0.47	1926.25	62.39 ± 2.60
Traditional coffee	Solid	2088.09 ± 84.27	1179.31 ± 126.10	25 401.52 ± 580.11	42.41 ± 2.21	156.33 ± 8.78	6.35 ± 0.46	15.96 ± 0.57	28 889.97	937.64 ± 5.97
	Brew	133.23 ± 0.83	41.79 ± 4.76	2014.91 ± 82.03	<2.28	0.87 ± 0.00	<3.74	<0.47	2190.80	48.09 ± 2.25
Instant coffee	Solid	3734.26 ± 198.40	1085.17 ± 49.67	16 823.59 ± 1000.45	41.91 ± 0.47	136.27 ± 0.87	15.25 ± 0.33	0.98 ± 0.01	21 837.43	737.70 ± 15.80
	Brew	62.99 ± 0.24	23.05 ± 2.75	840.07 ± 12.93	<2.28	<0.53	<3.74	<0.47	926.61	194.42 ± 1.86
Mean	–	9103.84	1645.80	1021.18	305.52	85.86	19.53	10.99	12 072.75	334.94

Mean values (n = 3) with standard deviations (SDs).
[a] Sum of minerals concentrations.

profile, similarities were observed between essential mineral concentrations of star anise and mate tea (herbal and spice matrices), as well as between all coffees. The differences in individual sample profiles may be mainly attributed to the chemical composition of the studied samples because they are different matrices, as expected.

In general, the highest total concentration of minerals in solid samples was found in mate tea (3218.18 mg kg^{-1}), followed by traditional coffee (28 889.97 mg kg^{-1}), espresso coffee 1 (28 162.45 mg kg^{-1}), espresso coffee 2 (24 941.89 mg kg^{-1}), instant coffee (21 837.43 mg kg^{-1}), and star anise (12 980.51 mg kg^{-1}). In the brew samples, the highest total concentration of minerals was found in traditional coffee (2190.80 mg kg^{-1}), followed by espresso coffee 1 (2104.03 mg kg^{-1}), espresso coffee 2 (1926.25 mg kg^{-1}), instant coffee (926.61 mg kg^{-1}), mate tea (486.97 mg kg^{-1}), and ground star anise (172.63 mg kg^{-1}). The intact star anise infusion (not milled) presented a total mineral concentration equal to 145.06 mg kg^{-1}.

An interesting fact was observed in relation to Ca concentration determined in star anise brews obtained from milled and from intact star anise—the intact star anise brew presented a high amount of Ca (86.40 mg kg^{-1}) when compared to the milled star anise brew (9.31 mg kg^{-1}). To confirm this result, Ca concentration was determined in both samples in a new test and our preliminary results were confirmed. Therefore, this finding clearly indicates that intact star anise brew presents a higher amount of Ca compared to ground star anise brew. This result might be related to the release of chelating agents when the star anise is ground, which probably take the form of strong complexes with Ca. It is well-known that lipids can reduce Ca absorption due to the formation of insoluble nonabsorbable calcium soap, especially in matrices with high lipid content [23]. Previous studies [5, 6] have shown that star anise has a high essential oil content in its composition, in addition to phenolic compounds, flavonoids, and carotenoids, which may also interact, especially with Ca, and decrease its extraction. Furthermore, recent studies [24] have shown that the phenolic compounds negatively influence the bioaccessibility of Ca in vegetable-based products. Although no phenolic identification has been carried out, these authors have found that, depending on the plant beverage formulation and its traditional phenolic composition, the bioaccessibility of minerals is affected. Therefore, in the present study, lipids and phenolic constituents from milled star anise may have interacted with Ca and have acted as chelating agents. This study is the first report on the star anise mineral composition; therefore, more samples should be included in further studies.

Total mineral leachable levels in the brews were calculated (table 8.2). Among the studied samples, it was observed that K, Ca, and Mg were leached into brews in highest degrees with rates between 0.67%–7.93%, 0.41%–7.30%, and 0.24%–6.38%, respectively. In contrast, it was verified that Mn and Fe were leached into brews in lowest degrees, with rates between 0.16%–0.72% and 0.52%–2.73%, respectively. For Zn and Cu, the values were not detectable, as shown in the table. The differences in the lixiviation rate between the solids and their brews might be related to the interactions of minerals with the chemical compounds present in the solid matrices, which may affect their extractions. This is not unusual and previous studies have already noted low rates of mineral released into traditional infusions [2, 25–27].

Table 8.2. Percentage (%) of essential minerals content extraction from solid samples to brews.

Sample	Mg	Ca	K	Mn	Fe
Star anise milled	1.37	0.79	1.76	0.16	ND
Intact star anise	0.24	7.30	0.67	ND	ND
Mate tea	1.00	0.41	2.49	0.72	2.73
Espresso coffee 1	5.38	3.05	7.85	ND	ND
Espresso coffee 2	5.17	4.65	8.13	ND	ND
Traditional coffee	6.38	3.54	7.93	0.56	ND
Instant coffee	1.69	2.12	4.99	0.21	0.52
Mean	3.03	3.12	4.83	0.41	1.63

ND = Not detectable.
LOD (limit of detection) = Mn (<0.53 mg kg^{-1}) and Fe (<2.28 mg kg^{-1}).

Traditional and espresso coffees exhibited a higher mineral extraction, whereas the opposite was observed for instant coffee. In addition, mineral extraction was higher for traditional and espressos coffees than for star anise, mate tea, and instant coffee samples, which might be related to the possible chemical differences between matrices, as well as brewing conditions, such as extraction time and temperature. In particular, for espresso coffee capsules, the pressure from the machines may have contributed to high mineral extraction because it is known that pressure significantly increases the extraction of constituents from foods. Although solid mate tea has presented the highest total mineral concentration, a low leachability rate was observed in its infusion. This might be related to the phenolic and flavonoid constituents that are widely present in mate tea that could interact with the minerals, form strong complexes and decrease their extraction [2, 25, 26, 28]. In general, these preliminary results show that the mineral composition and its leachability depend largely on the studied matrices.

The results obtained for TPCs are presented in table 8.1. Regarding solid samples, the highest TPC was found in traditional coffee (937.64 mg GAEq/L), followed by espresso coffee 2 (927.30 mg GAEq/L), espresso coffee 1 (923.85 mg GAEq/L), instant coffee (737.70 mg GAEq/L), mate tea (272.85 mg GAEq/L), and star anise (43.09 mg GAEq/L). In brew samples, the highest TPC was found in instant coffee (194.42 mg GAEq/L), followed by mate tea (108.07 mg GAEq/L), espresso coffee 2 (62.39 mg GAEq/L), espresso coffee 1 (61.36 mg GAEq/L), traditional coffee (48.09 mg GAEq/L), and star anise (26.37 mg GAEq/L). The brew from intact star anise presented a TPC of 11.15 mg GAEq/L. These levels are within the range previously reported for traditional infusions from different sources [3, 5, 7]. Among bioactive substances found in plant-based products such as coffee, mate tea, and star anise, phenolic compounds are the major contributors to their total antioxidant activity [5, 7].

Figure 8.1 illustrates the TPC leachable levels in the brews. Even though the star anise sample showed the lowest TPC among the solid samples, its brew from ground star anise presented the highest TPC (more than 60%). Meanwhile, unground star

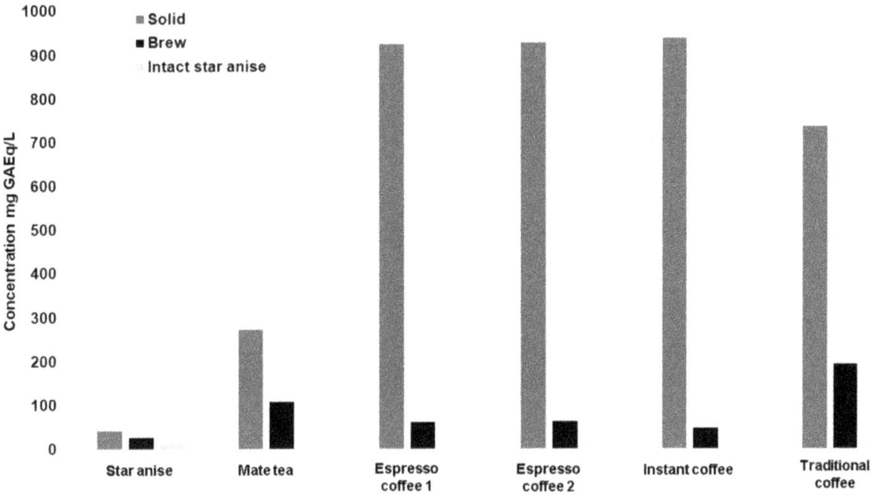

Figure 8.1. Comparison of TPC in the solid samples and their respective brews.

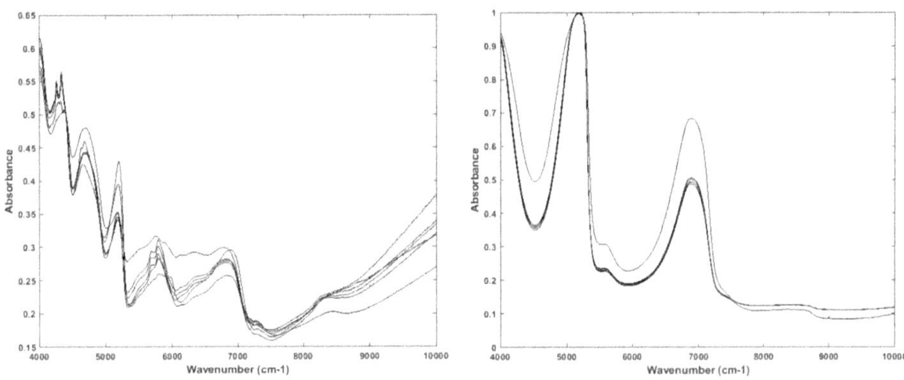

Figure 8.2. NIR spectra of the preprocessed solid samples (A) and their respective brews (B).

anise brew showed a lower leaching rate (less than 26%). Therefore, more phenolic compounds were extracted in the brew from ground star anise. This may have favorably affected Ca extraction in the brew sample from ground star anise, which corroborates our previous results. In relation to the TPC rate from the other brew samples, mate tea presented a rate of more than 39%, espresso coffee 2 equal to 7.73%, espresso coffee 1 of 6.64%, traditional coffee equal to 5.13%, and instant coffee more than 26%. The TPC leached levels are within the range reported in the literature for traditional infusions [3, 5, 7].

8.3.2 Qualitative evaluation of the NIR spectra and smartphone-based image

The preprocessed NIR spectra of the samples are shown in figure 8.2. As expected, there was a clear difference between solid samples and their brews. Moreover, a distinct NIR profile was observed between samples from different sources. These

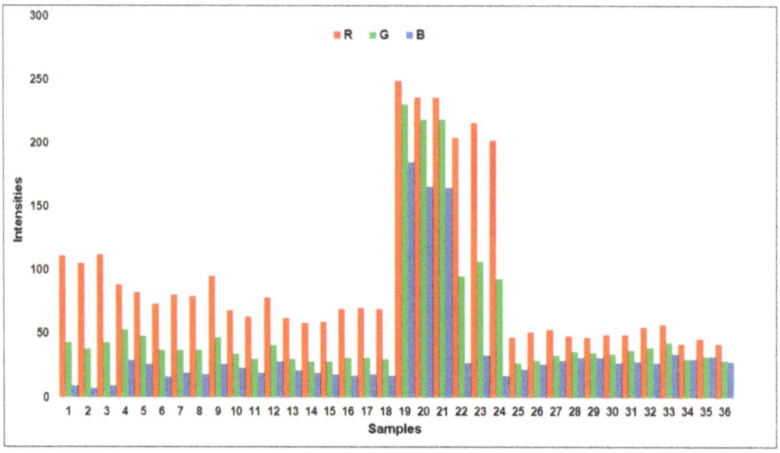

Figure 8.3. RGB mean values of each solid or brew sample. Solid samples and their respective replicates are in the following ranking: star anise (1–3), mate tea (4–6), instant coffee (7–9), espresso coffee 1 (10–12), espresso coffee 2 (13–15) and traditional coffee (16–18). Brew samples and their respective replicates are in the following ranking: star anise (19–21), mate tea (22–24), instant coffee (25–27), espresso coffee 1 (28–30), espresso coffee 2 (31–33), and traditional coffee (34–36).

signatures could be attributed to the changes in the absorption bands in the first, second, and third overtone regions referring to CH, OH, and NH functional groups of chemical compounds present in the samples [29]. It can also be observed that brew samples have a similar spectral profile, which can be associated with the OH functional group from the water used in the brewing preparation.

The results for smartphone-based image analysis are shown in figure 8.3. Overall, RGB intensities were similar for most samples. It was also verified that color intensities were more prominent in solid samples (from 1 to 18) in comparison to brew samples (from 19 to 36). Only star anise and mate tea brew samples showed a tendency to higher intensities, probably associated with the light tone of these samples compared to coffee tones (darker). The red intensity was highest for all samples, and the green and blue intensities were the smallest.

NIR spectroscopy and smartphone-based image analyzes showed interesting results from visual examination; however, due to the lack of selectivity of these techniques, it is difficult to draw conclusions on the importance of these analyzes for an overall evaluation of the samples. This indicates that a chemometric data treatment is required to better explore these results (see below).

8.3.3 Data fusion and PCA

The results from the experiments (i.e. TPC, mineral composition, NIR spectra, and RGB mean values) performed on the samples were merged via a low-level data fusion with a subsequent chemometric evaluation by PCA. This approach was tested for simultaneous evaluation of different infusion sources (solids and their brews) in an attempt to improve the interpretability of the results and to show the potential of these tools to verify patterns of similarity between samples, and also to show how the brew can compromise the mineral content.

The PCA provides information on the samples and variables through plots from scores and loadings (Fig.3A-F), respectively. In addition, it allows us to evaluate which part of each data set has a greater influence on the dispersion shown in the scores. The results obtained for data fusion coupled with PCA are discussed in detail in this section.

The first two principal components (PCs) were able to explain more than 95% of the variance in the fusion matrix (88.19% in PC1 and 7.30% in PC2). Although PC2 showed a lower explained variance than PC1, previous studies have shown that often relevant information from complex samples is extracted from PCs with the lowest explained variances [17, 30, 31]. In general, solid fractions were discriminated from their respective infusions in PC1, probably because the differences between them were greater, highlighting that it comes from different systems (i.e. solid or brew). Meanwhile, PC2 discriminated the samples, showing the relation between them. This made it possible to verify the similarities between the samples coming from different systems.

Concerning PC1 scores shown in figure 8.4(A), all of the solid samples were discriminated from their brews (and respective replicates), which is probably associated with the large differences between solid and liquid fractions in these samples. The solid fractions presented negative scores (cluster one), while their respective brews presented positive scores (cluster two). The evaluation of PC1 loadings shows how the variables influenced the separation of the scores. For NIR (figure 8.4(B)), all of the spectral region is important for clusters observed in scores. However, since brews are liquid fractions, the OH functional group from water molecules certainly was the most significant vibration for the distinction between solid and their brews. Moreover, other constituents (e.g. phenolic compounds, carbohydrates, lipids, and caffeine) are present in lower concentrations in these studied matrices, which may also influence the NIR vibrations or overtones.

In relation to the other analyzes (i.e. TPC, mineral composition, and RGB intensitics) performed in both samples, it was observed that all of the variables contributed to the separation of the scores (figure 8.4(C)). However, green and blue intensities were the main variables responsible for discriminating brew samples, while essential minerals, TPC, and red intensity were important for discriminating solid samples. A more detailed observation of the scores (figure 8.4(A)) in combination to the loadings (figure 8.4(C)) provides evidence that mineral concentration, TPC, and red intensity are higher in the solid samples than in the liquid fractions, which corroborates our previous results (table 8.1 and figure 8.3). Therefore, this discrepancy may have been attributed to the differences between the composition of solid and liquid fractions.

Even though PC1 plots show good results, PC2 plots exhibited the most interesting findings. The patterns of similarity between infusions from different sources were verified through PC2. Observing the distribution of the scores (figure 8.4(D)) combined with the PC2 loading plots (figure 8.4(E) and (F)) on the

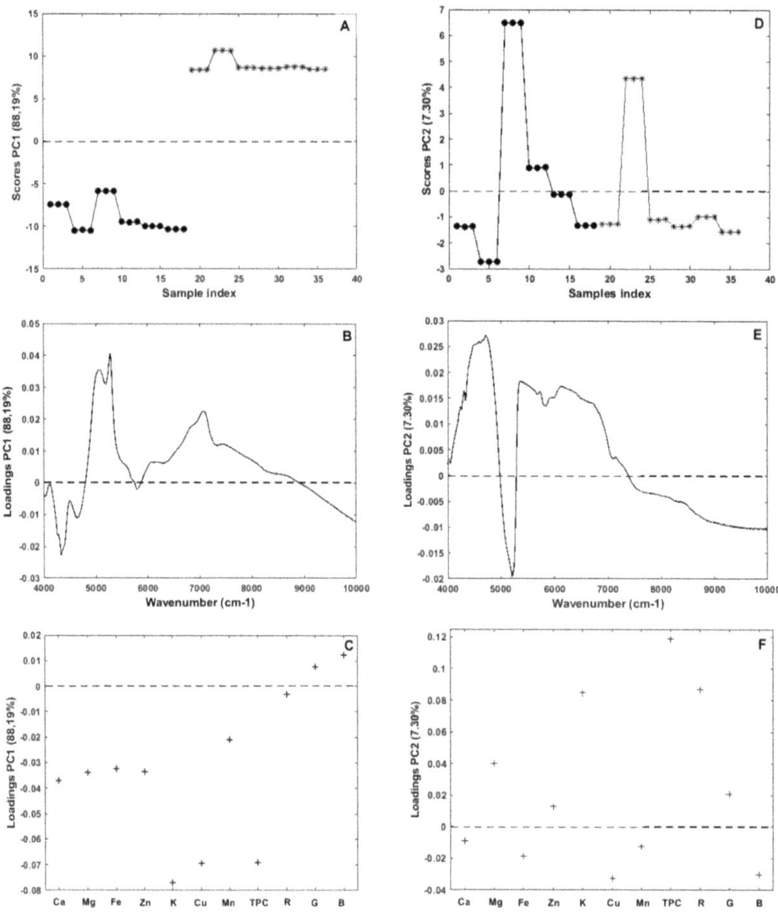

Figure 8.4. PCA scatter plots from PC1 and PC2 of all of the samples and their respective replicates. (A) PC1 scores; (B) PC1 NIR loadings; (C) PC1 mineral composition, TPC, and RGB loadings; (D) PC2 scores; (E) PC2 NIR loadings; (F) PC2 mineral composition, TPC, and RGB loadings. Different symbols represent solid (●) and brew (*) samples. Solid samples and their respective replicates are in the following ranking: star anise (1–3), mate tea (4–6), instant coffee (7–9), espresso coffee 1 (10–12), espresso coffee 2 (13–15), and traditional coffee (16–18). Brew samples and their respective replicates are in the following ranking: star anise (19–21), mate tea (22–24), instant coffee (25–27), espresso coffee 1 (28–30), espresso coffee 2 (31–33), and traditional coffee (34–36).

negative side from PC2, overall, the most significant results are that solid samples from star anise and traditional coffee presented a similarity with brew samples from star anise, instant, espresso, and traditional coffee, associated with Ca, Fe, Cu, and Mn levels. These results suggest that even with differences on the matrix sources, their behavior is similar (solid and liquid fractions). This is associated with mineral composition, which is verified by the proposed method.

Although the total concentrations of minerals in the studied samples are different, this does not have a large impact on final mineral proportion, depending on the matrices. This may arise from many variables, such as the chemical matrix or the slightly similar infusion (i.e. hot water extraction). Further studies may reveal interactions between minerals and specific compounds in these matrices, which will help to clarify these results.

In addition, other observations were verified to a lower degree in PC2 scores, which highlighted slight similarities between infusion fractions from different sources. The solid samples from star anise, mate tea, espresso coffee 2, and traditional coffee, as well as brew samples from star anise, instant coffee, traditional coffee, and espresso coffees 1 and 2, exhibited negative scores, and therefore there is a tendency of similarity between these samples. In contrast, solid samples from instant coffee and espresso coffee 1, and the brew sample from mate tea showed positive scores. These relationships can be attributed to the variations in PC2 loadings plots, which highlight the similarities and differences between the samples (figures 8.4(E) and (F)).

Although no previous study has described a simultaneous evaluation of mate tea, coffees, star anise, and their brews, the data fusion associated with PCA proposed in the present study proved to be a useful and powerful tool that is able to evaluate these products and show the similarities between the studied samples. For the industry, the development of chemometric models from the combination of large data sets can help professionals to better understand their processes and the characteristics of their products.

8.4 Conclusion

This study combined different data sets to evaluate and study the relationships between powders of coffees, mate tea, star anise, and their respective infusions. The most significant results obtained through data fusion with chemometrics showed that the proposed method is useful to show unknown relationships between samples coming from different matrices. Their similar behavior, both in powder and in their respective infusions, can be related to the similar compositional characteristics in both products. Even though more samples should be included in the study to consider differences between suppliers and batches, our approach has shown promising results and the possibility to use this method to simultaneously evaluate foods from different sources.

Due to advances in analytical instrumentation, the development of integrated approaches to monitor processes and their products in industrial processing routines has recently increased. However, simultaneous measurements are only useful if both forms of information are valuable. There is still a lack of studies of the potential of these approaches to evaluate distinct products, considering the context of the food industry. Our approach is based on data fusion with subsequent chemometric evaluation and seems to be attractive for quality control purposes involving the acquisition of multiple data sets with analytical complexity in a single set of samples. This method could be used in the future as a basis to implement mathematical models based on chemometrics to qualitatively evaluate unknown samples.

Acknowledgments

This work was supported by the São Paulo Research Foundation (FAPESP) (grant #2019/21062-0 and grant #2022/04068-8, and process #2022/03268-3), Coordenação de Aperfeiçoamento de Pessoal de Nível Superior (CAPES) - Finance Code 001, Conselho Nacional de Desenvolvimento Científico e Tecnológico (CNPq) (process 306606/2020-8, 402441/2022-2, and process 310982/2020-0).

References

[1] International Coffee Organization 2016 The current state of the global coffee trade | CoffeeTradeStats (http://ico.org/monthly_coffee_trade_stats.asp)

[2] Giulian R, Dos Santos C E I, De Moraes Shubeita S, Da Silva L M, Dias J F and Yoneama M L 2007 Elemental characterization of commercial mate tea leaves (*Ilex paraguariensis* A. St.-Hil.) before and after hot water infusion using ion beam techniques *J. Agric. Food Chem.* **55** 741–6

[3] Heck C I and Mejia E G D 2007 Yerba mate tea (*Ilex paraguariensis*): a comprehensive review on chemistry, health implications, and technological considerations *J. Food Sci.* **72**

[4] Bhadra S, Mukherjee P K, Kumar N S and Bandyopadhyay A 2011 Anticholinesterase activity of standardized extract of Illicium verum Hook. f. fruits *Fitoterapia* **82** 342–6

[5] Aly S E, Sabry B A, Shaheen M S and Hathout A S 2016 Assessment of antimycotoxigenic and antioxidant activity of star anise (*Illicium verum*) *in vitro J. Saudi Soc. Agric. Sci.* **15** 20–7

[6] Wang G W, Hu W T, Huang B K and Qin L P 2011 *Illicium verum*: a review on its botany, traditional use, chemistry and pharmacology *J. Ethnopharmacol.* **136** 10–20

[7] Ferruzzi M G 2010 The influence of beverage composition on delivery of phenolic compounds from coffee and tea *Physiol. Behav.* **100** 33–41

[8] Orlando E A, Rebellato A P, Silva J G S, Andrade G C and Pallone J A L 2020 Sodium in different processed and packaged foods: method validation and an estimative on the consumption *Food Res. Int.* **129**

[9] Pohl P, Stelmach E, Welna M and Szymczycha-Madeja A 2013 Determination of the elemental composition of coffee using instrumental methods *Food Anal. Methods* **6** 598–613

[10] Baqueta M R, Coqueiro A, Março P H and Valderrama P 2020 Quality control parameters in the roasted coffee industry: a proposal by using MicroNIR spectroscopy and multivariate calibration *Food Anal. Methods* **13** 50–60

[11] Baqueta M R, Coqueiro A and Valderrama P 2019 Brazilian coffee blends: a simple and fast method by near-infrared spectroscopy for the determination of the sensory attributes elicited in professional coffee cupping *J. Food Sci.* **84**

[12] Pallone J A L, dos E T, Caramês S and Alamar P D 2018 Green analytical chemistry applied in food analysis: alternative techniques *Curr. Opin. Food Sci.* **22** 115–21

[13] Rateni G, Dario P and Cavallo F 2017 Smartphone-based food diagnostic technologies: a review *Sensors (Switzerland)* **17**

[14] Schwolow S, Gerhardt N, Rohn S and Weller P 2019 Data fusion of GC-IMS data and FT-MIR spectra for the authentication of olive oils and honeys—is it worth to go the extra mile *Anal. Bioanal. Chem.* **411** 6005–19

[15] Banerjee M B, Roy R B, Tudu B, Bandyopadhyay R and Bhattacharyya N 2019 Black tea classification employing feature fusion of e-nose and e-tongue responses *J. Food Eng.* **244** 55–63

[16] Di Rosa A R, Leone F, Cheli F and Chiofalo V 2017 Fusion of electronic nose, electronic tongue and computer vision for animal source food authentication and quality assessment—a review *J. Food Eng.* **210** 62–75

[17] Gonçalves T R, Rosa L N, Torquato A S, da Silva L F O, Março P H, Gomes S T M, Matsushita M and Valderrama P 2020 Assessment of brazilian monovarietal olive oil in two different package systems by using data fusion and chemometrics *Food Anal. Methods* **13** 86–96

[18] Barbosa C D *et al* 2020 Data fusion of UPLC data, NIR spectra and physicochemical parameters with chemometrics as an alternative to evaluating kombucha fermentation *LWT—Food Sci. Technol.* **133** 109875

[19] Ou B, Hampsch-Woodill M and Prior R L 2001 Development and validation of an improved oxygen radical absorbance capacity assay using fluorescein as the fluorescent probe *J. Agric. Food Chem.* **49** 4619–26

[20] Singleton V L and Rossi Jr J A 1965 Colorimetry of total phenolics with phosphomolybdic-phosphotungstic acid reagents *Am. J. Enol. Vitic.* **16** 144–58

[21] Geladi P, MacDougall D and Martens H 1985 Linearization and scatter-correction for near-infrared reflectance spectra of meat *Appl. Spectrosc.* **39** 491–500

[22] Savitzky A and Golay M J E 1964 Smoothing and differentiation of data by simplified least squares procedures *Anal. Chem.* **36** 1627–39

[23] Lorieau L, Le Roux L, Gaucheron F, Ligneul A, Hazart E, Dupont D and Floury J 2018 Bioaccessibility of four calcium sources in different whey-based dairy matrices assessed by *in vitro* digestion *Food Chem.* **245** 454–62

[24] Silva J G S, Rebellato A P, dos E T, Caramês S, Greiner R and Pallone J A L 2020 *In vitro* digestion effect on mineral bioaccessibility and antioxidant bioactive compounds of plant-based beverages *Food Res. Int.* **130** 108993

[25] Erdemir U S 2018 Contribution of tea (*Camellia sinensis* L.) to recommended daily intake of Mg, Mn, and Fe: an *in vitro* bioaccessibility assessment *J. Food Compos. Anal.* **69** 71–7

[26] Stelmach E, Pohl P and Szymczycha-Madeja A 2013 The suitability of the simplified method of the analysis of coffee infusions on the content of Ca, Cu, Fe, Mg, Mn and Zn and the study of the effect of preparation conditions on the leachability of elements into the coffee brew *Food Chem.* **141** 1956–61

[27] Wróbel K, Wróbel K and Urbina E M C 2000 Determination of total aluminum, chromium, copper, iron, manganese, and nickel and their fractions leached to the infusions of black tea, green tea, *Hibiscus sabdariffa*, and *Ilex paraguariensis* (Mate) by ETA-AAS *Biol. Trace Elem. Res.* **78** 271–80

[28] Ashu R and Chandravanshi B S 2011 Concentration levels of metals in commercially available ethiopian roasted coffee powders and their infusions *Bull. Chem. Soc. Ethiop.* **25** 11–24

[29] Jerry Workman J and Weyer L 2012 *Weyer, Practical Guide and Spectral Atlas for Interpretive Near-Infrared Spectroscopy* (Boca Raton, FL: CRC Press) (https://books.google.com.br/books?hl=pt-BR&lr=&id=i_eWjO5pb28C&oi=fnd&pg=PP1&ots=VsJRFVMDQG&sig=HAIgRGcaar-Qr3PwUmLvcphgalcM&redir_esc=y#v=onepage&q&f=false)

[30] Moreira I and Scarminio I S 2013 Chemometric discrimination of genetically modified *Coffea arabica* cultivars using spectroscopic and chromatographic fingerprints *Talanta* **107** 416–22

[31] de Almeida M M C, Francisco C R L, de Oliveira A, de Campos S S, Bilck A P, Fuchs R H B, Gonçalves O H, Velderrama P, Genena A K and Leimann F V 2018 Textural, color, hygroscopic, lipid oxidation, and sensory properties of cookies containing free and microencapsulated chia oil *Food Bioprocess Technol.* **11** 926–39

IOP Publishing

Advances in Spectroscopic Analysis of Food and Drink

Ashutosh Kumar Shukla

Chapter 9

Near-infrared spectroscopy for fruit/fruit juice quality analysis: theory and application

Rasool Khodabakhshian

9.1 Introduction

This chapter will give a a brief overview of Raman spectroscopy applications in analyzing fruits and vegetables. The discovery of near-infrared energy is attributed to Herschel in the nineteenth century; however, the first industrial application began in the 1950s. Near-infrared spectroscopy (NIRS) was initially only used as a complementary unit to other optical equipment that used wavelengths such as ultraviolet (UV), visible (Vis) or mid-infrared (MIR) spectrometers. In the 1980s, a stand-alone NIRS system was developed, but its application was mostly focused on chemical analysis. With the innovation of optical fibers in the mid-1980s and the development of monochromatic detectors in the early-1990s, NIRS became a powerful instrument for scientific studies. This optical technique can be used in a number of science fields, such as physics, chemistry, medicine, agriculture, and food.

The use of near-infrared spectrometers has proven to be a rapid and often non-destructive method for measuring the composition of biological materials for many products. This method is not new, even in the field of other foods, and it is an official method to determine the protein content of wheat. Nowadays, many institutions, universities, and research centers are famous for using this technology to determine the quality of agricultural produce. For example, the Food Research Institute (NFRI), Tsukuba has been a leading infrared spectroscopic research institute in Japan since 1970, and the Central Institute for Postharvest Engineering (CIPHET) is considered as a pioneer to conduct research in the field of food quality measurement using NIRS in India. In Iran, in addition to dedicated spectroscopic laboratories located in universities (including Sharif University of Technology, University of Tehran, Ferdowsi University of Mashhad and Khajeh Nasir Toosi), research institutes such as Bo Ali Research Institute of Mashhad University of Medical Sciences, Laser and Plasma Research Institute of Shahid Beheshti University and

doi:10.1088/978-0-7503-5573-5ch9

© IOP Publishing Ltd 2023

Laser, Optics and Photonics Research Institute of Amirkabir University are pioneering the supply of this technology.

9.2 Theoretical principles

NIRS uses the near-infrared range of the electromagnetic spectrum (from about 700–2500 nm) and is based on the vibration of molecular compounds. As a result, molar absorptivity in the near-infrared range is usually very small, but an advantage is that near-infrared can usually penetrate much deeper into a sample than MIR radiation. NIRS is therefore not only a particularly sensitive technique but it can also be very useful in examining bulk materials with minimal or no preparation. The bands of molecular compounds observed in the near-infrared are typically very broad and result in complex spectra. Therefore, it can be difficult to assign spectral characteristics to specific chemical compounds. Multivariate calibration methods (such as principal component analysis or partial least squares (PLS) analysis, which are discussed in this chapter) are often used for reflection, transmission, and absorption phenomena in order to extract the desired chemical information. The careful development of a set of calibration samples and the application of multivariate calibration methods for near-infrared analytical methods are essential. In the following, the theoretical aspects of near-infrared are discussed:

9.2.1 Electromagnetic properties of radiation

The infrared wave is a part of electromagnetic spectrum that is considered as a simple harmonic wave. The electric and magnetic properties of these waves are interconnected and interact with matter and create a spectrum. A simple harmonic motion has the characteristic of a sine wave, specified by equation (9.1):

$$y = A \sin \theta \tag{9.1}$$

where y is a displacement with a maximum value of A, and θ is an angle between zero and 2π radians.

Now consider a circular path with radius A that point P moves on with a uniform angular velocity ω radians/second (figure 9.1); P defines an angle $\theta = \omega t$ radians after t seconds passing Q, and so its vertical displacement is

$$y = A \sin \omega t \tag{9.2}$$

Equation (9.2) is shown graphically on the right-hand side of figure 9.1. P returns to point Q after $2\pi/\omega$ seconds and completes one cycle in 1 s and the pattern repeats $\omega/2\pi$. This parameter is called the frequency (ν) of the wave and the basic equation (9.3) of the wave motion can be written as follows:

$$y = A \sin 2\pi\nu t \tag{9.3}$$

The another property of the wave is wavelength (λ), which means the distance traveled in a complete cycle. Equation (9.3) requires expression in terms of displacement variation with distance instead of time This will be done by submitting $t = lc$, where l is the displacement covered by the wave in time t at velocity c.

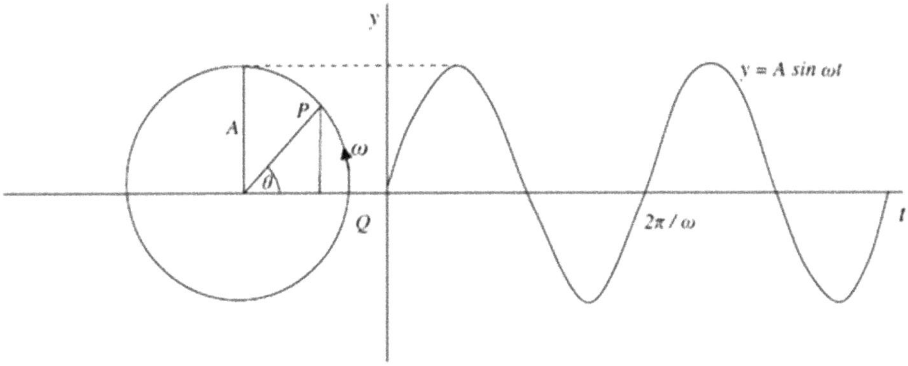

Figure 9.1. Schematic representation of the wave of an electromagnetic radiation [1]. Reprinted by permission from Springer Nature Customer Service Centre GmbH: Springer Nature, Springer eBook, Copyright © 2010, Springer-Verlag Berlin Heidelberg.

The velocity c is an universal constant and is identified as light velocity in vacuum. So, the wavelength can be specified as follows:

$$\lambda = c/\nu \tag{9.4}$$

The wavenumber ($\bar{\nu}$) in physics expresses the spatial frequency of a wave in centimeters, which is measured in cycles per unit distance or radians per unit distance, and is another way of expressing the nature of electromagnetic radiation. Spectroscopy experts describe the position of an infrared absorption band in the wavenumber period. As mentioned, it is proportional to the frequency ($\nu = c\bar{\nu}$) and is related to the energy variations involved in the transition between different vibrational states.

The wave model has been used to explain many parameters of electromagnetic radiation, but it cannot be used to describe phenomena related to energy absorption or transfer. So, it is essential to observe electromagnetic radiation as a stream of discrete particles, termed photons, with an energy proportional to the radiation frequency (equation (9.4)).

9.2.2 Vibrational properties of molecules

9.2.2.1 The harmonic oscillator

Mechanical Model: To recognize the harmonic oscillation of a molecule in vibration mode, assume a mass m at the end of a spring that is fixed at the other end. The force of gravity is constant and consequently the only effect is the balance point. Disturbance of the mass along the axis of the spring results in a motion described by Hooke's law. This explains that the restoring force F exerted by the spring is proportional to the displacement y (the displacement traveled from the balance point).

$$F = -ky \tag{9.5}$$

where k is the force constant. The acceleration of a mass from the balance point is:

$$a = \frac{d^2y}{dt^2} \tag{9.6}$$

Using Newton's second law in motion, F = ma:

$$\frac{md^2y}{dt^2} = -ky \tag{9.7}$$

A differential solution of equation (9.7) is:

$$y = A \sin \alpha t \tag{9.8}$$

where α is the positive square root of k/m. After a period of motion, y will return to its initial value and the sine wave will repeat each time and αt will increase by $2\pi\nu$. Therefore, it is concluded as follows:

$$\nu = \frac{1}{2\pi}\sqrt{\frac{k}{m}} \tag{9.9}$$

By putting $\alpha = \sqrt{k/m} = 2\pi\nu$ in equation (9.8), we reach equation (9.3). Electromagnetic waves and mechanical oscillators can therefore be described in the same conditions with the first approximation, and the importance of spectroscopic measurements is placed in the relationship between the frequency of radiant energy and the frequencies of molecular movements.

Equation (9.9) can be modified to explains the behavior of a system including two masses, m_1 and m_2, attached to a spring, by substituting the reduced mass $\mu = (m_1m_2)/(m_1 + m_2)$ for m:

$$\nu = \frac{1}{2\pi}\sqrt{\frac{k}{\mu}} \tag{9.10}$$

Any system involving more than two masses follows the same or more complex equations. So, the vibration of a chemical bond can be assumed to resemble the behavior of a spring when m_1 and m_2 are the masses of two atoms, and k is the force constant for the chemical bond. By using this simple mechanical model, it is possible to describe many infrared spectral observations. For example, it has been experimentally determined that the compound containing a carbonyl group (carbon–oxygen double bond) C = O has an infrared band in the wavenumber region of 1500–1900 cm^{-1} [2].

By putting $\nu = c\bar{\nu}$ and placing the approximate values of 1×10^3 N m^{-1} for the force constant of a double bond, 2×10^{-26} and 2.7×10^{-26} kg, respectively, for the masses of carbon and oxygen, and 3×10^8 m s^{-1} for the velocity of light in equation (9.10) we have:

$$\bar{\nu} = \frac{1}{2 \times 3.14 \times 3 \times 10^8}\sqrt{\frac{1 \times 10^3(2 + 2.7)\times 10^{-26}}{2 \times 2.7 \times 10^{-52}}} = 1.565 \times 10^5\ \mathrm{m}^{-1},\ \text{or}\ 1565\ \mathrm{cm}^{-1}$$

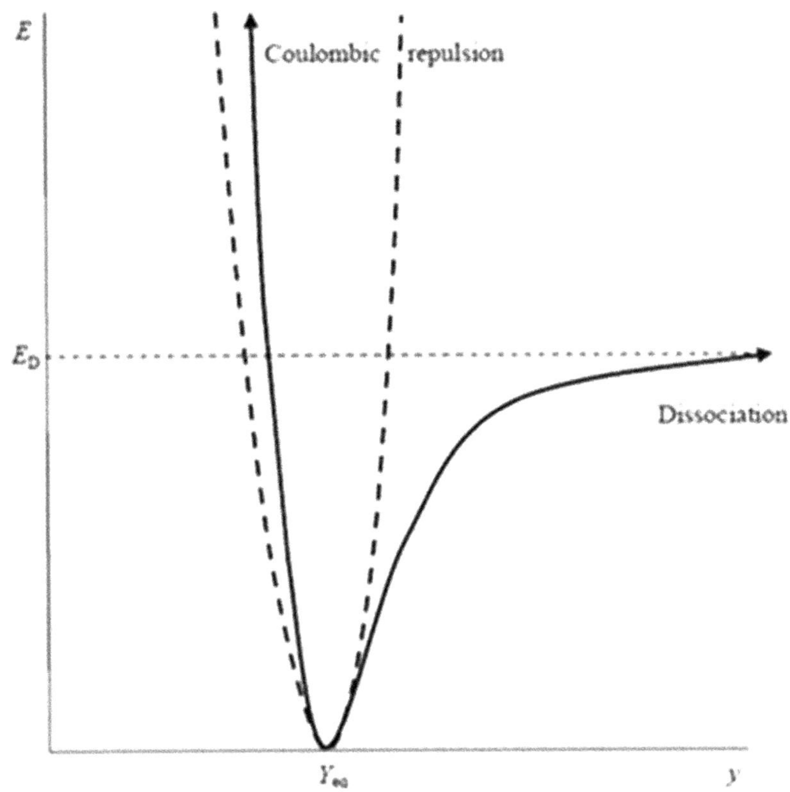

Figure 9.2. The potential energy curve of a diatomic molecule under simple harmonic and balanced motion (dotted curve) and anharmonic vibration (solid curve) [3]. John Wiley & Sons. Copyright © 2009 John Wiley & Sons, Inc.

According to the mass and spring model, it is clear that the energy of the system is converted from potential energy to kinetic energy under a cycle. Figure 9.2 shows the potential energy diagram for a harmonic oscillator as a dotted curve.

At equilibrium point, the potential energy can be assumed to be zero, but when the spring is compressed or stretched by a low energy value dE:

$$dE = -Fdy \tag{9.11}$$

By combining equations (9.5) and (9.11), we have:

$$dE = -kydy \tag{9.12}$$

By integrating from $y = 0$ to y:

$$\int_0^E dE = k\int_0^y ydy \quad E = k\frac{y^2}{2} \tag{9.13}$$

The kinetic energy is equal to zero and the total energy is equal to the potential energy at the return point of the movement according to the maximum amplitude A.

Quantum Mechanics Model: At the end of the twentieth century there was not enough experimental data to support the assumption that matter had the ability to continuously absorb energy.

In 1900, Max Planck proposed that the energy of a oscillator is discontinuous and changes in its content occur through transitions between discrete energy states created by the absorption or emission of discrete packets of energy, called quantum. This idea became known as quantum theory, and energy levels are defined by integers called quantum numbers.

A transition between energy levels E_1 and E_2 occurs when the energy range $\Delta E = E_2 - E_1$ is absorbed or emitted by the system (figure 9.3). Planck also suggested that this energy is in the form of electromagnetic radiation, and the frequency of radiation is related to the energy change ΔE and is expressed by the following equation:

$$\Delta E = h\nu \tag{9.14}$$

where h is Planck's universal constant. Equation (9.14) means that if a radiation beam containing a wide range of frequencies hits a molecule in the E_1 energy state, then energy from the beam will be absorbed and a transition to the E_2 energy state will occur. A detector placed to collect the radiation after its interaction with the molecule shows that the intensity of the radiation decreases at the frequency $\nu = \Delta E/h$, and all other frequencies remain unchanged and the absorption action is provided (figure 9.3). In practice, the number of energy levels for a molecule is infinite, many transition processes are possible, and so a spectrum of the simplest molecule will be very complex.

A molecule in space has different forms of energy, such as vibrational energy because of the periodic displacement of its atoms from their balance position and rotational energy owing to the rotation of the object around its center of gravity. Absorption of infrared radiation is largely limited to molecular types because there is an energy difference between vibrational and rotational states. The energy required to change the rotational state is much smaller than the vibrational state and the rotational absorption bands may be absorbed only in the position of gases. So, to study the infrared spectrum of liquid and solid samples, only the vibrational motion should be considered.

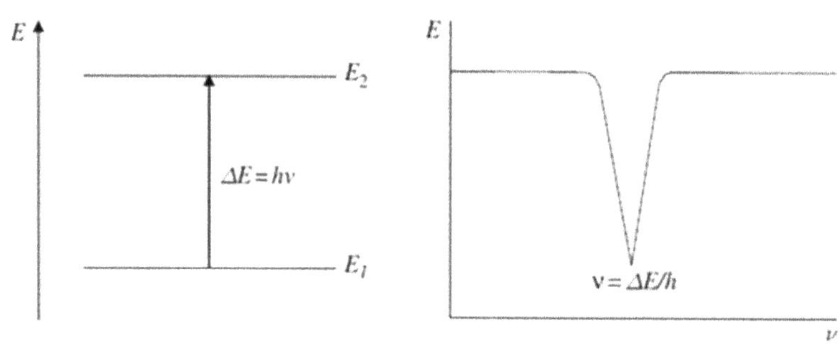

Figure 9.3. Visualization the quantum theory of absorption and radiation.

Vibrational energies, similar other molecular energies, are quantized (determining the properties and qualities of something by theory and formula) and allowed vibrational energies for each specific system are obtained by solving a series of partial differential equations, known as mechanical wave equations. The solution of these equations is obtained by assuming a simple balanced oscillator for the energy levels, as follows:

$$E = (\upsilon + 0.5)h\nu \tag{9.15}$$

where υ is the vibrational quantum number (0, 1, 2, ...). It can be found from equation (9.15) that the lowest level of vibrational energy E_0 at $\upsilon = 0$ is equal to $h\nu/2$. So, since atoms can never be completely at rest relative to each other, a molecule can never have zero vibrational energy. E depends only on the strength of the chemical bond and atomic masses. The prediction of E is the fundamental difference between classical and wave mechanical methods for molecular vibrations. Therefore, jumping to the first excited state ($\upsilon = 1$) needs the absorption of radiation energy, $(3h\nu/2) -(h\nu/2) = h\nu$. The radiation frequency (ν) that illuminates the limits of these changes is equal to the bond vibration frequency defined by equation (9.10), so:

$$\Delta E = \frac{h}{\nu}\sqrt{\frac{k}{\mu}} \tag{9.16}$$

From equation (9.15), it is found that E given by equation (9.15) is the energy associated with the transition between each pair of adjacent surfaces.

9.2.2.2 The anharmonic oscillator

The quantum mechanical behavior of a harmonic oscillator describes the observed infrared absorption bands with respect to the fundamental modes of molecular vibration, but it does not clarify the presence of overtone bands in the near-infrared. These bands occur as a result of transition processes (changing and passing from one state to another state) when $\Delta\upsilon$ is equal to ±2, ±3, etc, and so there is a prohibition and a violation of the rule. This irregularity is due to the natural and real nature of molecules, which do not strictly follow the laws of simple balanced motion and real bonds: although they have an elastic nature, they do not strictly follow Hooke's law. When two atoms approach each other, the Coulombic repulsion (due to Coulombic forces) between the two atomic nuclei causes a rapid increase in the potential energy compared to the predictions of the harmonic approximation. When the interatomic distance reaches the point where dissociation occurs, the energy levels reach a level state (figure 9.2). From the dotted curve in figure 9.2, it can be seen that the success of the steps of the harmonic model lies in the fact that the two curves are approximately similar at low potential energies. An empirical function with respect to the Morse potential energy (a simple model for the potential energy of a diatomic molecule by the famous physicist Philip Morse) fits the solid curve in figure 9.2 with a suitable estimate, as follows:

$$E = E_d(1 - e^{-\alpha y})^2 \tag{9.17}$$

where α is a constant for a particular molecule and E_d is the dissociation energy. Equation (9.17) is used to solve the wave mechanical equation. The solution for harmonic oscillator is equal to equation (9.18):

$$E = (v + 0.5)h\nu - (v + 0.5)^2 h\nu x - (v + 0.5)^3 h\nu x' - \ldots \quad (9.18)$$

where x, x', ... are small and positive harmonic constants that decrease in value and size. For small values of v, the third term and subsequent terms in equation (9.18) may be ignored and we have:

$$E \approx (v + 0.5)h\nu - (v + 0.5)^2 h\nu x \approx h\nu[1 - x(v + 0.5)](v + 0.5) \quad (9.19)$$

Equation (9.15) expresses the same energy level as equation (9.19), by substituting ν we have:

$$\nu' = \nu[1 - x(v + 0.5)] \quad (9.20)$$

As a result, the anharmonic oscillator behaves similarly to the harmonic oscillator with an oscillation frequency that steadily decreases as ν increases. E_0 is now equal to $(1–0.5x)(h\,\nu/2)$ and the energy associated with the transition from state v to $v+\Delta v$ can be expressed as equation (9.21):

$$\Delta E = h\nu[1 - (2v + \Delta v + 1)x] \quad (9.21)$$

9.2.2.3 Chemical assignments of near-infrared bands

Defining a number of momentum coordinates in order to describe the system is necessary to calculate the number of possible vibration modes for a polyatomic molecule. A space is defined using three coordinates. So, to define n points in space, $3n$ coordinates are needed. Three samples of these momentum coordinates define the translational motion of the entire molecule and the other three samples specify the same rotational motion. This exhibit $3n$–6 coordinates to define the interatomic vibrations. Since rotation around the bond axis is not possible, a linear molecule needs only two coordinate systems to define rotational motion. So, $3n$–5 modes in this situation are possible. For a very simple molecule, it is possible to define the nature and also the number of vibration modes (table 9.1).

Table 9.1. The approximate wavelength ranges of the infrared region [1]. Reprinted by permission from Springer Nature Customer Service Centre GmbH: Springer Nature, Springer eBook, Copyright © 2010, Springer-Verlag Berlin Heidelberg.

Region	Characteristic transitions	Wavelength range (nm)	Wavenumber (cm^{-1})
Near-infrared (NIR)	Overtones combinations	700–2500	14 300–4000
Mid-infrared (MIR)	Fundamental vibrations	$2500–5 \times 10^4$	4000–200
Far-infrared (FIR)	Rotations	$5 \times 10^4–10^6$	200–10

Vibration is categorized in either of two modes: stretching or bending. If there is a continuous change in the interatomic distance along the bond axis between two atoms, then the vibration is called stretching, which can occur symmetrically in a triatomic group of AX_2 atoms or asymmetrically when they vibrate in different directions. Vibration including a change in bond angle is known as bending and is categorized into four types: scissoring, rocking, wagging, and twisting. These cases are also known as symmetrical inside the plane of deformation, asymmetrical inside the plane of deformation, symmetrical outside the plane of deformation, and asymmetrical outside the plane of deformation.

Each of these vibrational modes gives rise to visible orton compounds in the near-infrared range. However, the intensity of such bands depends on the degree of anharmonicity. For example, a bond involving hydrogen, the lightest atom, vibrates with a large amplitude when under stretching, and thus the motion deviates noticeably from harmonic. As a result, almost all the observed absorption bands due to hydrogen stretching vibration overtones include AHy basic groups or compounds that include stretching and bending vibration modes of such groups in the infrared region. For more information on this field, we refer the interested reader to books on molecular geometry in chemistry and books on analytical chemistry.

9.2.2.4 Interpretation of the near-infrared spectrum

Although the summary of the chemical assignments for the available near-infrared bands signifies only some important near-infrared bands, the degree of spectral complexity is apparent. A library of 328 infrared spectra of pure chemical compounds and food and agricultural products is available in a thesis by Williams and Norris (1987) [4]. Another large library was developed over time with the participation of various researchers as a public reference and can be obtained from the following Internet address: www.spectraonline.com.

9.2.3 Near-infrared imaging

Besides chemical allocations (assignments) of different near-infrared bands for the analysis of food and other substances, near-infrared imaging is also applied for the chemical imaging of samples. Chemical imaging has an analytical capability to create a visual image of the simultaneous measurement of the spatial dimension of the spectrum (x and y values) and time information. This method is often applied for any solid or gelatin samples and has applications in chemistry, biology, medicine, pharmacy, food industry, biotechnology, agriculture, and industry. Near-infrared imaging is also referred to as hyperspectral, spectral, or multispectral imaging.

9.3 Near-infrared spectroscopy equipment

The previous sections dealt with what NIR spectroscopy is and how it is used for quality assessment. The user usually has an increasingly difficult task in determining the most appropriate device for use in the laboratory or in a practical environment.

The instrumentation used in non-destructive spectrometry to measure absorption or reflection includes a light source, wavelength selector, holder, detector, and a computer. Nowadays, these components are connected to each other through an optical fiber or a single string, except for the computer. The wavelength selector and detector are combined in the spectrometer. The parts of these components of NIRS are of great importance, and therefore further details will be given in the following subsections.

9.3.1 Near-infrared spectroscopy modes

There are several different methods for measuring NIR rays and choosing the optimal method can play a significant role in the ability of predictive models. In general, NIR measurement methods can be divided into three reflection, transmission, and interactance methods. In the reflective method, the light source and the detector are placed in the same direction and at an angle of 45 degrees to each other to prevent normal reflection. In the transmission method, the transmitted beam that has passed through the tissue of the sample is measured. The transmission method is divided into two methods based on the arrangement of the beam source and the detector: full transmission measurement method and semi-transmission measurement method. In the full transmission measurement system, the beam source, sample, and detector are placed on both sides of the sample and in the same direction, while in the semi-transmission method the direction of the beam source and the detector are perpendicular to each other. Finally, in the contrast method, which is generally performed using bifurcated optical fiber, the light source and the detector are placed parallel to each other and in such a way that the rays caused by normal reflection cannot enter the detector directly.

9.3.2 Source of light

Conventional incandescent or quartz halogen lamps are the most widely-used sources of broadband near-infrared radiation for analytical applications. Quartz halogen lamps are the most preferred source of near-infrared energy due to their widespread use for home lighting. Using the word energy in the case of NIRS is more appropriate compared to the word light because near-infrared energy is not visible to the human eye, and therefore is not light. Light emitting diodes (LEDs) are the second most popular source of near-infrared energy.

The way in which the light is spectrally balanced or selected defines the basis of the optical activity of a near-infrared spectrometer. All such equipment is classified in three groups: dispersive, interferometric, and non-thermal. The devices in the first two groups of thermal radiation use the broadband produced by an incandescent lamp filament, while the third group includes non-thermal or cold sources.

9.3.3 Near-infrared spectrometers

Near-infrared spectrometers are classified according to the methods of wavelength isolation (wavelength selection). Wavelength isolators (wavelength selectors) fall broadly into two categories: (a) discrete value and (b) full spectrum devices. Full spectrum spectrometers, also known as spectrum scanning equipment, operate with data at equal intervals, i.e., at constant intervals, along the full range from 700 to

2500 nm. The distance can be any unique number, such as 1.5, 2, 2.2 nm and so on. Discrete-quantity spectrometers are further classified by the technology used to produce narrow-bandwidth wavelength bands.

9.3.4 Infrared detectors

Infrared detectors are divided into two types: photon (optical) detectors including photoconductors (light-conducting) and photodiodes (light diodes), and thermal detectors. The response characteristics of some useful near-infrared detectors are shown in table 9.2. Responsivity (R), noise equivalent power (NEP) and specific detectivity (D^*) are terms that are commonly used to define detectors. NEP is defined as the power input required to provide an output noise voltage equal to the root mean square of the noise voltage, while the inverse of NEP is the detectivity D. When D is expressed in a standard form for a 1 cm^2 surface detector and a hertz bandwidth, it is referred to as a special detection capability and is expressed as D^*.

9.3.5 Optical fibers

The use of optical fibers as a means of delivering and transmitting near-infrared energy and information is extremely important. There are many situations due to temperature, vibration, space, or explosion risks (especially in the production line where the work environment is not suitable for sensitive equipment) in which the ability of optical fibers to deliver and transmit near-infrared energy is a major advantage. The principle of internal reflection governs the science of optical fibers. It supplies all features of light transmission along optically transparent fibers using the phenomenon of total internal reflection. The transparent medium may be silica glass, plastic, or any material like these.

Unlike natural reflections from a metal or dielectric surface, which may be as high as 99%, total internal reflection is very efficient and typically better than 99.999% per reflection. Even with a few hundred or a few thousand reflections per meter, significant ranges of energy can be transmitted over suitable distances. Fiber attenuation is measured in inverse decibel kilometers or inverse decibel meters. The attenuation of three fibers suitable for the near-infrared is shown in figure 9.4.

The best chalcogenide fibers cover the wavelength range from 4 to 11 μm, but data transmission is limited to distances of a few meters. As can be found from this figure, the silica glass is ideal for the range below 2 μm. Zirconium fluoride is expensive in the current situation and offers very weak attenuations, less than 1 dB km^{-1}.

9.4 Statistical data analysis

9.4.1 Data preprocessing

Preprocessing has been shown to improve the stability of the model [6]. However, it is important to take into account that preprocessing does not always improve the results and an unsuitable method can even have a negative effect on a model's accuracy [7]. Often, studies utilize more than one preprocessing method, and the performance of different methods is compared [8, 9]. However, in most cases, only

Table 9.2. Infrared detectors and their characteristics. Reprinted from [5], Copyright (1970), with permission from Academic Press, Inc. Published by Elsevier Ltd.

Types			Detectors	Spectral response (μm)	Operating temperature (K)	D^* (cm Hz 1/2 /W)
Thermal	Thermocouple, thermopile		—	Depends on window material	300	$D^*\ (\lambda,10,1) = 6 \times 10^8$
	Bolometer		—		300	$D^*\ (\lambda,10,1) = 1 \times 10^8$
	Pneumatic cell		Golay cell, condenser-microphone		300	$D^*\ (\lambda,10,1) = 1 \times 10^9$
	Pyroelectric detector		PZT, TGS, $LiTaO^3$		300	$D^*\ (\lambda,10,1) = 2 \times 10^8$
Photon detector	Intrinsic	Photoconductive	Si	0.85–1	295	$D^* \approx 4 \times 10^{12}$
			PbS	1–3.6	300	$D^*(500,600,1) = 1 \times 10^9$
			PbSe	1.5–5.8	300	$D^*(500,600,1) = 1 \times 10^8$
			InSb	2–6	213	$D^*(500,1200,1) = 2 \times 10^9$
			HgCdTe	2–16	77	$D^*(500,1200,1) = 2 \times 10^{10}$
		Photovoltaic	Ge	0.8–1.8	300	$D^*(\lambda_p) = 1 \times 10^{11}$
			InGaAs	0.7–1.7	300	$D^*(\lambda_p) = 5 \times 10^{12}$
			Ex. InGaAs	1.2–2.55	253	$D^*(\lambda_p) = 2 \times 10^{11}$
			InAs	1–3.1	77	$D^*(500,1200,1) = 1 \times 10^{10}$
			InSb	1–5.5	77	$D^*(500,1200,1) = 2 \times 10^{10}$
			HgCdTe	2–16	77	$D^*(500,1200,1) = 1 \times 10^{10}$
	Extrinsic		Ge:Au	1–10	77	$D^*(500,900,1) = 1 \times 10^{11}$
			Ge:Hg	2–14	4.2	$D^*(500,900,1) = 8 \times 10^9$
			Ge:Cu	2–30	4.2	$D^*(500,900,1) = 5 \times 10^9$
			Ge:Zn	2–40	4.2	$D^*(500,900,1) = 5 \times 10^9$
			Si:Ga	1–17	4.2	$D^*(500,900,1) = 5 \times 10^9$
			Si:As	1–23	4.2	$D^*(500,900,1) = 5 \times 10^9$

$D^*$is normally expressed in format $D^*(a, b, c)$, where a is the temperature (K) or wavelength (μm) of radiant energy, b is the chopping frequency, and c is the bandwidth. Subscript p denotes peak wavelength in this table.

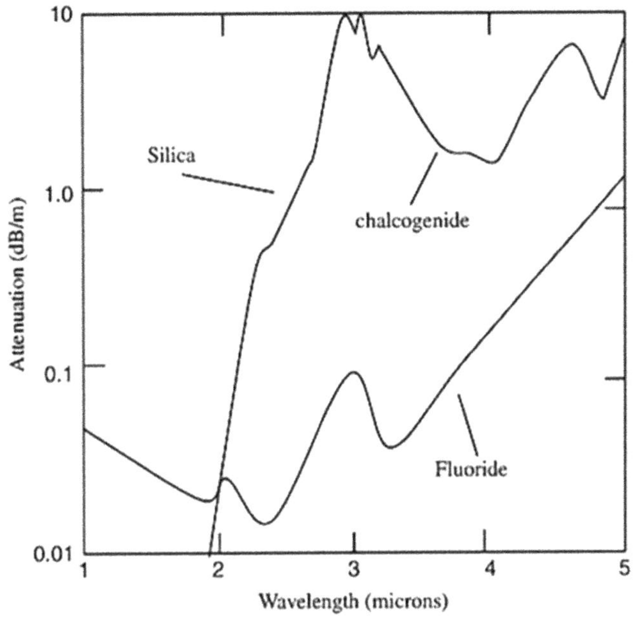

Figure 9.4. An estimate of near-infrared transmission in optical fibers [1]. Reprinted by permission from Springer Nature Customer Service Centre GmbH: Springer Nature, Springer eBook, Copyright © 2010, Springer-Verlag Berlin Heidelberg.

one preprocessing method, which can be a single method or a combination of methods, is reported and used for processing, modeling, and classification purposes [10–12]. In the absence of a standard selection methodology, preprocessing techniques are chosen based on the researchers' experience or the previous studies done in their field. Currently, there are many preprocessing methods, some of which are: spectroscopic transformations, normalization, derivation, standard normal variable transformation (SNV), multiplicative scatter correction (MSC), and new methods such as wavelet transformation [13].

9.4.2 Data description

Keeping the data simple, describing it with useful abbreviations, and often displaying the inherent data structures visually with appropriate data graphics are large parts of multivariate analysis. As an example, the desired data may be the values of the constant parameters monitored in an industrial process at different locations or the measured variables (e.g., temperature, pressure, etc) from a series of synthesis of organic compounds. The purposes of describing univariate and multivariate data can be very diverse, e.g., determination of simple averages and standard deviations, etc, as well as correlation of applied regression models. There are a number of multivariate techniques, such as Principal Component Analysis (PCA), Principal Components Regression (PCR), PLS, and Multiple Linear Regression (MLR).

9.4.3 Regression and prediction

Regression has been widely used in the fields of science and technology. It is an approach to relate two sets of variables together. In addition, it involves predicting one or more Y variables based on a well-chosen set of suitable X variables, where X should generally include more than three variables. Prediction means determining Y values for new X variables based on a pre-estimated (calibrated) X–Y model. As a result, it relies only on the new X data. Although many types of analysis methods are available in statistics books, only the important and usable ones in near-infrared analysis and modeling, such as PLS, principal component regression (PCR), and multiple linear regressions (MLRs), are briefly described in the following subsections.

9.4.3.1 Partial least squares

PLS is a two-way modeling method (bilinear modeling) in which the information in the original X data is visualized on a small number of hidden (latent) variables, called PLS components. The Y information is actively used in the estimation of the latent variables to ensure that the first components are those that are most suitable for predicting the Y variables. The interpretation of the relationship between data X and data Y is then simplified into a relationship focused on the smallest possible number of components. By plotting the basic PLS components, one can see the dependencies between X and Y variables, as well as mutual relationships between Y data and X data. There are two versions of the PLS algorithm: PLS1 deals with only one response (dependent) variable at a time and PLS2 simultaneously deals with multiple responses.

9.4.3.2 Principal component regression

PCR is a method that is suitable in the same situations and conditions as PLS. It is a two-step method. First, a principal component analysis is performed on X variables. The original components are then applied as predictors in an MLR method.

9.4.3.3 Multiple linear regressions

MLR is a method of relating changes in a response variable (Y variable) to changes in several X variables for descriptive or predictive objectives. An important assumption for this method is that the X variables are linearly independent, meaning that there is no linear relationship between the X variables. Their changes or each other are not considered, so the variance of X is not significant there. Difficulties can arise with respect to exact or approximate collinearity when the X variables carry routine information.

9.4.4 Classification

Classification, or what is called discrimination, of samples separates data groups. Suppose that you have a large number of measurements of oranges and after analyzing them you have decided to group the measurements into two groups, or when a food inspector collects milk samples from different sources to detect adulterated samples. In this situation numerical values have to be assigned to these two groups to make a digital decision. Therefore, classification requires an advanced class description. Many grouping techniques are available and the following subsections will briefly review some of them.

9.4.4.1 Neural networks

Among the classification techniques, it is likely that the neural network method is the most widely known. The main advantage of neural network techniques is that they are all-purpose, they can handle problems with multiple parameters, and they are able to group objects appropriately even when the distribution of objects in N-dimensional space is complex. The disadvantage of neural network methods is that they are extremely slow, especially in the training phase but also in the execution phase. Another notable weakness of neural networks is that it is very difficult to determine how the network makes decisions. As a result, it is difficult to determine which of the sample features used are important and useful for grouping and which are worthless.

9.4.4.2 Nearest-neighbor classifiers

A very simple classification technique can be based on a nearest-neighbor approach. In this technique, it is possible to find the closest object from the trained set to an object being grouped in a multi-dimensional feature space. Since the concept of neighborhood means the closest and with high commonality, it is very likely that the neighborhood of an object is very similar to the object being grouped, and therefore has the same group as the object being grouped. The nearest-neighbor technique has many advantages, such as ease of use. If the features are carefully selected, this method can also provide very good results. There are, however, serious disadvantages of nearest-neighbor technique. First, like neural networks, they do not simplify the distribution of objects in the parametric space into a comprehensible set of parameters. The most serious disadvantage of the nearest-neighbor technique is that it is very sensitive to the presence of unrelated parameters.

9.4.4.3 Decision trees

The decision tree technique has also been used to solve many problems. In the parallel decision tree technique, a binary tree is built in which a single parameter is compared with a number of fixed values in each node. If the feature value exceeds the threshold, it belongs to the right-hand branch of the tree; if this value is smaller, then the left-hand branch is followed. After a series of these tests, we reach a leaf node of the tree that has assigned all objects to a specific group. These are called parallel axis trees because they are associated with the partitioning of the parameters. Parallel decision trees are typically much faster in the construction (training) phase than neural network techniques, and they also lead to faster executioner space into a set of hyperplanes that are parallel to all feature axes except the one under test. Their weakness is that they are not flexible in modeling parameter space distributions with complex distributions, such as any of the two techniques of neural networks and nearest neighbor. In fact, even simple shapes can cause problems with these methods.

9.4.5 Validation

Validation of a model means that the performance of the model is tested on an unknown set of samples that were not used in the calibration. This data set is called the test set. In the following, a brief overview is provided to introduce important validation issues that should be considered when determining a multivariate calibration.

9.4.5.1 Test set validation

The approach introduced above using a raw data set for validation is called a test set validation. The discussed approach in test set validation allows the model to predict *Y* values and then compare these predicted values with actual and known *Y* values (which were not used in the modeling and prediction process) to compare independently. In general, the predicted *Y* values are known as Y_{pred} and the actual *Y* values are known as Y_{ref}. An ideal test set condition should have a sufficiently large number of training set measurements for both *X* and *Y*. This data set is then used for model calibration. Now, an independent and secondary sampling of the target population is carried out to produce a test set that is only used to validate the model by comparing Y_{pred} and Y_{ref}.

9.4.5.2 Cross-validation

There is no better validation than test set validation. One should always try to validate through the test set. However, there are definitely situations and conditions in which this procedure is clearly not possible, e.g., when the measurement of Y values is too expensive, acceptance is compromised, or when the preparation of samples is too difficult. For this situation, there is a suitable alternative method called cross-validation. In the most favorable conditions, cross-validation can be almost as good as a test set validation, but it can never be a substitute for a proper test set validation.

9.4.5.3 Leverage corrected validation

This method is a fast and overly optimistic (in dirty terms) validation method. This method also uses the same calibration set to validate the model, but at this stage with leverage correction. It is clear that this method is a questionable validation process because it depends entirely on the quality of the correction applied. In addition, the results of this method are too optimistic. However, during early modeling periods when validation is not really on the agenda, this method can be useful because it saves time.

9.5 Applications of near-infrared spectroscopy for fruit and fruit juice

Several studies have been conducted regarding the use of NIRS in the quality assessment of fruit and fruit juice. In table 9.3, an overview of the applications of near-infrared spectroscopy for the measurement of soluble solids content in fruits and

Table 9.3. An overview of applications of near-infrared spectroscopy to quality assessment of fruit and fruit juice.

Product type	Spectroscopy mode	Spectral range (nm)	References
Apple	Reflectance	380–1650	[14]
	Reflectance	380–2000	[15]
	Interactance	500–1100	[16]
	Transmittance	650–950	[17]
	Reflectance	400–2500	[18]
	Transmittance	650–950	[19]
	Reflectance	400–1700	[20]
	Interactance	812–2357	[21]
	Reflectance	400–1000	[22]
Orange	Transmittance	636–1236	[23]
	Transmittance	400–980	[24]
Apricot	Interactance	600–2500	[25]
	Reflectance	650–1200	[26]
Cherry	Reflectance	400–1700	[25]
	Interactance	729–975	[27]
Grape	Transmittance	650–1100	[28]
Kiwifruit	Reflectance-transmittance	300–1100	[29]
	Reflectance	400–1100	[30]
Tangerine	Reflectance	800–1100	[31]
	Interactance	300–1100	[32]
	Transmittance	350–2500	[33]
Mango	Reflectance	1200–2400	[34]
	Interactance	700–1100	[35]
Watermelon	Reflectance	700–1100	[36]
	Interactance	300–1150	[37]
Nectar	Interactance	735–932	[38]
Peach	Transmittance	400–1000	[39]
	Reflectance	800–2500	[40]
Pear	Reflectance	780–1700	[41]
	Reflectance	425–1000	[42]
Pineapple	Reflectance	650–1000	[43]
	Reflectance	400–1700	[44]
Plum	Interactance	306–1150	[38]
	Interactance	500–1010	[45]
Banana	Reflectance	400–2500	[46]
Pomegranate	Reflectance	400–1100	[47]
	Transmittance	400–1000	[48]

vegetables is presented. It is clear from this table that while suitable calibration models have been obtained for dry matter, it is difficult to predict acidity based on near-infrared spectrum. The review of the studies presented in this table shows that while suitable calibration models have been obtained for dry matter, it is difficult to predict acidity based on the near-infrared spectrum. In addition, the concentration of acids in many fruits and vegetables is significantly lower than that of sugars, and they are probably too small to affect the near-infrared spectrum. Meanwhile, water absorption bands cover the spectrum of fruits and vegetables and it is highly likely that it does not have the ability to measure minor compounds well. It is obvious when examining the correlation of the concentration of such partial compounds such as sugar content that the calibration results may seem reasonable, but the robustness of the model should be checked for application because it is an indirect method.

Today, industry is the leader in the development of online infrared spectroscopy systems to measure the quality of fruits. The first line-on-line sorting system with a near-infrared sensor in reflective mode was introduced in 1989 by Mitsui Minig to grade peaches based on soluble solids content. Since then, several manufacturers have followed this procedure and sorting lines with near-infrared sensors are commercially available today, including:

- Awaeta (IQA, http://www.aweta.nl),
- Greefa (iFA, http://www.greefa.nl),
- Mitsui-Kinzoku (http://www.mitsuikinzoku.co.jp),
- Sacmi (F5, http://www.sacmi.it),
- TasteMark (http://www.taste-technologies.com),
- Color Vision Systems (Insight, http://www.cvs.com.au)

Unfortunately, there are no accurate scientific statistics regarding the accuracy of these systems.

References

[1] Jha S N 2010 Near infrared spectroscopy ed S N Jha *Nondestructive Evaluation of Food Quality: Theory and Practice* (Berlin: Springer) pp 141–212

[2] Osborne B G, Fearn T and Hindle P H 1983 Practical NIR spectroscopy in food and beverage analysis

[3] Nakamoto K 1997 *Infrared and Raman Spectra of Inorganic and Coordination Compounds* 5th edn (New York: Wiley)

[4] Williams P and Norris K 1987 *Near Infrared Technology in the Agricultural and Food Industries* (St Paul, MN: American Association of Cereal Chemists Inc.) pp 247–90

[5] Levinstein H 1970 Chapter 1 Characterization of infrared detectors *Semiconductors and Semimetals* (Amsterdam: Elsevier) pp 3–12

[6] Li S, Xing B, Lin D, Yi H and Shao Q 2020 Rapid detection of saffron (*Crocus sativus* L.) adulterated with lotus stamens and corn stigmas by near-infrared spectroscopy and chemometrics *Ind. Crops Prod.* **152** 112539

[7] Mishra P, Rutledge D N, Roger J-M, Wali K and Khan H A 2021 Chemometric pre-processing can negatively affect the performance of near-infrared spectroscopy models for fruit quality prediction *Talanta* **229** 122303

[8] Brendel R, Schwolow S, Rohn S and Weller P 2020 Comparison of PLSR, MCR-ALS and Kernel-PLSR for the quantification of allergenic fragrance compounds in complex cosmetic products based on nonlinear 2D GC-IMS data *Chemometr. Intell. Lab. Syst.* **205** 104128

[9] Gerhardt N, Schwolow S, Rohn S, Pérez-Cacho P R, Galán-Soldevilla H, Arce L and Weller P 2019 Quality assessment of olive oils based on temperature-ramped HS-GC-IMS and sensory evaluation: comparison of different processing approaches by LDA, kNN, and SVM *Food Chem.* **278** 720–8

[10] Jamwal R, Amit Kumari S, Balan B, Kelly S, Cannavan A and Singh D K 2021 Rapid and non-destructive approach for the detection of fried mustard oil adulteration in pure mustard oil via ATR-FTIR spectroscopy-chemometrics *Spectrochim. Acta, Part* A **244** 118822

[11] Obeidat S M, Hammoudeh A Y and Alomary A A 2018 Application of FTIR spectroscopy for assessment of green coffee beans according to their origin *J. Appl. Spectrosc.* **84** 1051–5

[12] Santos D I, Neiva Correia M J, Mateus M M, Saraiva J A, Vicente A A and Moldão M 2019 Fourier transform infrared (FT-IR) spectroscopy as a possible rapid tool to evaluate abiotic stress effects on pineapple by-products *Appl. Sci.* **9** 4141

[13] Nicolai B M, Beullens K, Bobelyn E, Peirs A, Saeys W, Theron K I and Lammertyn J 2007 Nondestructive measurement of fruit and vegetable quality by means of NIR spectroscopy: a review *Postharvest Biol. Technol.* **46** 99–118

[14] Lammertyn J, Nieolai B O and Smedt V D *et al* 1998 Nondestructive measurement of acidity, soluble solids and firmness of jonagold apples using NIR-spectroscopy *Trans ASAE* **41** 1089–94

[15] Ventura M, Jager A and Putter H *et al* 1998 Non-destructive determination of soluble solids in apple fruit by near infrared spectroscopy *Postharvest Biol. Technol.* **14** 21–7

[16] Peirs A, Ooms K, Lammertyn J and Nicolaı B 2001 Prediction of the optimal picking date of different apple cultivars by means of VIS/NIR spectroscopy *Postharvest Biol. Technol.* **21** 189–99

[17] McGlone V A, Jordan R B and Martinsen P J 2002 Vis/NIR estimation at harvest of pre- and post-storage quality indices for 'Royal Gala' apple *Postharvest Biol. Technol.* **25** 135–44

[18] McGlone V A and Martinsen P J 2004 Transmission measurements on intact apples moving at high speed *J. Near Infrared Spectrosc.* **12** 37–43

[19] McGlone V A, Martinsen P J, Clark C J and Jordan R B 2005 On-line detection of Brownheart in Braeburn apples using near infrared transmission measurements *Postharvest Biol. Technol.* **37** 142–51

[20] Xing J, Bravo C, Moshou D, Ramon H and De Baerdemaeker J 2006 Bruise detection on 'Golden Delicious' apples by vis/NIR spectroscopy . *Comp. Electron. Agric.* **52** 11–20

[21] Liu Y, Ying Y, Yu H and Fu X 2006 Comparison of the HPLC method and FT-NIR analysis for quantification of glucose, fructose and sucrose in intact apple fruits *J. Agric. Food Chem.* **54** 2810–5

[22] Pourdarbani R, Sabzi S and Arribas J I 2021 Nondestructive estimation of three apple fruit properties at various ripening levels with optimal vis-NIR spectral wavelength regression data *Heliyon* **7** e07942

[23] Lee K, Kim G, Kang S, Son J, Choi D and Choi K 2004 Measurement of sugar content in citrus using near infrared transmittance *Key Eng. Mater.* **270–273** 1014–9

[24] Song J, Li G, Yang X, Liu X and Xie L 2020 Rapid analysis of soluble solid content in navel orange based on visible-near infrared spectroscopy combined with a swarm intelligence optimization method *Spectrochim. Acta, Part* A **228** 117815

[25] Carlini P, Massantini R and Mencarelli F 2000 Vis-NIR measurement of soluble solids in cherry and apricot by PLS regression and wavelength selection *J. Agric. Food Chem.* **48** 5236–42

[26] Camps C and Christen D 2009 Non-destructive assessment of apricot fruit quality by portable visible-near infrared spectroscopy *LWT Food Sci. Technol.* **42** 1125–31

[27] Escribano S, Biasi W V, Lerud R, Slaughter D C and Mitcham E J 2017 Non-destructive prediction of soluble solids and dry matter content using NIR spectroscopy and its relationship with sensory quality in sweet cherries *Postharvest Biol. Technol.* **128** 112–20

[28] Herrera J, Guesalaga A and Agosin E 2003 Shortwave-near infrared spectroscopy for non-destructive determination of maturity of wine grapes *Meas. Sci. Technol.* **14** 689–97

[29] Schaare P N and Fraser D G 2000 Comparison of reflectance, interactance and transmission modes of visible-near infrared spectroscopy for measuring internal properties of kiwifruit (*Actinidia chinensis*) *Postharvest Biol. Technol.* **20** 175–84

[30] Moghimi A, Aghkhani M H, Sazgarnia A and Sarmad M 2010 Vis/NIR spectroscopy and chemometrics for the prediction of soluble solids content and acidity (pH) of kiwifruit *Biosyst. Eng* **106** 295–302

[31] Kawano S, Fujiwara T and Iwamoto M 1993 Nondestructive determination of sugar content in satsuma mandarin using near infrared (NIR) transmittance *J. Jpn. Soc. Hortic. Sci.* **62** 465–70

[32] Walsh K B, Golic M and Greensill C V 2004 Sorting of fruit using near infrared spectroscopy: application to a range of fruit and vegetables for soluble solids and dry matter content *J. Near Infrared Spectrosc.* **12** 141–8

[33] Gomez H A, He Y and Pereira A G 2006 Non-destructive measurement of acidity, soluble solids and firmness of Satsuma mandarin using Vis/NIRspectroscopy techniques *J. Food Eng.* **77** 313–9

[34] Schmilovitch Z, Mizrach A, Hoffman A, Egozi H and Fuchs Y 2000 Determination of mango physiological indices by near-infrared spectrometry *Postharvest Biol. Technol.* **19** 245–52

[35] Saranwong S, Sornsrivichai J and Kawano S 2004 Prediction of ripe-stage eating quality of mango fruit from its harvest quality measured non-destructively by near infrared spectroscopy *Postharvest Biol. Technol.* **31** 137–45

[36] Guthrie J A, Wedding B and Walsh K B 1998 Robustness of NIR calibrations for soluble solids in intact melon and pineapple *J. Near Infrared Spectrosc.* **6** 259–65

[37] Long R L and Walsh K B 2006 Limitations to the measurement of intact melon total soluble solids using near infrared spectroscopy *Aust. J. Agric. Res.* **57** 403–10

[38] Golic M and Walsh K B 2006 Robustness of calibration models based on near infrared spectroscopy for the in-line grading of stone fruit for total soluble solids content *Anal. Chim. Acta* **555** 286–91

[39] Carlomagno G, Capozzo L, Attolico G and Distante A 2004 Non-destructive grading of peaches by near-infrared spectrometry *Infrared Phys. Technol.* **46** 23–9

[40] Ying Y B, Liu Y D, Wang J P, Fu X P and Li Y B 2005 Fourier transform near-infrared determination of total soluble solids and available acid in intact peaches *Trans. ASAE* **48** 229–34

[41] Nicolaı B M, Verlinden B E, Desmet M, Saevels S, Theron K, Cubeddu R, Pifferi A and Torricelli A 2007 Time-resolved and continuous wave NIR reflectance spectroscopy to predict firmness and soluble solids content of Conference pears *Postharvest Biol. Technol.*

[42] Khodabakhshian R and Emadi B 2017 Application of Vis/SNIR hyperspectral imaging in ripeness classification of pear *Int. J. Food Prop.* **20** S3149–63

[43] Chia K S, Abdul Rahim H and Abdul Rahim R 2012 Prediction of soluble solids content of pineapple via non-invasive low cost visible and shortwave near infrared spectroscopy and artificial neural network *Biosystems Eng.* **113** 158–65

[44] Xu S, Ren J, Lu H, Wang X, Sun X and Liang X 2022 Nondestructive detection and grading of flesh translucency in pineapples with visible and near-infrared spectroscopy *Postharvest Biol. Technol.* **192** 112029

[45] Li M, Lv W, Zhao R, Guo H, Liu J and Han D 2017 Non-destructive assessment of quality parameters in 'Friar' plums during low temperature storage using visible/near infrared spectroscopy *Food Control* **73** 1334–41

[46] Tarkosova J and Copikova J 2000 Determination of carbohydrate content in bananas during ripening and storage by near infrared spectroscopy *J. Near Infrared Spectrosc.* **8** 21–6

[47] Khodabakhshian R, Emadi B, Khojastehpour M and Golzarian M R 2016 Carob moth, *Ectomyelois ceratoniae*, detection in pomegranate using visible/near infrared spectroscopy *Comput. Electron. Agric.* **129** 9–14

[48] Khodabakhshian R, Emadi B, Khojastehpour M and Golzarian M R 2019 A comparative study of reflectance and transmittance modes of Vis/NIR spectroscopy used in determining internal quality attributes in pomegranate fruits *J. Food Meas. Charact.* **13** 3130–9

Chapter 10

The use of elemental analysisbased on atomic spectroscopy for quality assurance of fruit

Agustina A M B Hastuti, Anjar Windarsih and Abdul Rohman

A number of elements are essential for the development and the quality of fruits, such as macronutrients, micronutrients, or trace elements. The elemental requirements to get optimum quality and yield depend on the type of the fruit. Although elemental concentrations strongly affect fruit yield and quality, there is no standardized method to reflect elemental adequacy. The elemental contents found in juice fruits could be used as indicative parameters during the evaluation of the quality of the fruit juice. The technique used for the analysis of elemental concentration is generally dictated by the number of elements analyzed, the number of samples, and the concentration of analyte. The determination of elemental concentrations in fruits is mostly performed using atomic absorption spectroscopy (AAS), inductively couple plasma-atomic emission spectroscopy (ICP-AES), or inductively couple plasma-mass spectrometry (ICP-MS). In addition, stable isotope data coupled with multi-element analysis could be used as an analytical tool for authentication studies, especially for the authentication of the geographical origins of fruit.

10.1 Introduction

The quality of fruit is often described by the external appearance to address the consumers' requirements. The appearance, size, shape, and color of the fruits are the first traits that determine the consumers choice of a certain fruit, alongside internal factors such as taste, acidity, texture, and nutritional value. While not directly affecting the consumers' decision, elemental content is one of the parameters that determine fruit quality. A deficient or an excess of an element may impair the nutritional adequacy of the fruit, resulting in lesser quality and yield. For example, nitrogen (N) is a major element that constitutes proteins and other cellular molecules [1]. Although the addition of N in the soil through fertilizer induces plant growth, an excess of N reduces apple quality due

doi:10.1088/978-0-7503-5573-5ch10 © IOP Publishing Ltd 2023

to the impairment of other elemental contents in the fruit [2]. Micro-elements, such as copper (Cu) and iron (Fe), are involved in many enzymatic processes and cellular signaling. Moreover, the concentration of elements can directly affect the fruit taste, such as in the case of potassium (K) that determines fruit acidity [1]. Meanwhile, hazardous levels of certain elements, such as Cu, chromium (Cr), and cadmium (Cd), can indicate low quality fruit and potential contamination from the environment [3].

The importance of elemental concentrations for the optimum quality of fruit and the preferred nutrition by the consumers allow elemental concentrations in fruit potential to be used to indicate overall quality of fruit. The deviation from optimum percentage (DOP) index is quite popular to demonstrate an adequacy of element in fruit [4]. A positive DOP indicates that the element is present in fruit in excess, while a negative DOP means that the elemental concentration in the fruit is lower than the standard. However, the quality assurance for fruit based on elemental concentration is still not widely applied in the industry. The elemental analysis is also important for the authentication studies of beverage products, including fruit juices [5].

The majority of the elemental concentrations in fruits have been determined using atomic spectroscopy. Atomic spectroscopy, including AAS, inductively coupled plasma-atomic emission spectroscopy (ICP-AES, also called as inductively coupled plasma-optical emission spectroscopy/ICP-OES), and inductively coupled plasma-mass spectrometry (ICP-MS), is the common method for elemental analysis. This method is usually chosen depending on the number of elements analyzed and the concentration. Some of the major elements were determined by simple methods specialized for certain elements, such as the Kjeldahl method to determine nitrogen (N) concentration and colorimetric analysis to determine phosphorus (P) concentration [6, 7]. In addition, stable isotope data coupled with multi-element analysis have been used as analytical tools for authentication studies, especially the geographical origins of fruit authentication [8]. Several studies have been conducted to propose the trace elements of isotope ratios as chemical markers during fruit authentication. Most of these studies were obviously devoted to the classification of high-quality fruits, such as typical productions labeled with Protected Designation of Origin (PDO), Protected Geographical Indication (PGI), or Traditional Specialty Guaranteed (TSG) [9].

10.2 Atomic spectroscopy

Atomic spectroscopy techniques are preferred for the determination of elemental concentration in fruit for a faster and high throughput analysis. Prior to instrumental analysis, samples are usually subjected to a digestion step to liberate elements contained in the samples. During digestion, the organic material present in the sample is mostly converted into N_2 and CO_2. Organic materials can affect the analytical performance by hindering material transfer into the detector, thus increasing carry-over in the tubing and reducing sensitivity. Wet digestion by adding high concentrated acid, such as HNO_3, HCl, H_2SO_4, and/or HF, with heating at a relatively low temperature (~100°C) is widely used. This technique is considered to be more efficient and can prevent excessive sample loss. A more efficient wet digestion method is available with the assistance of microwave irradiation.

Microwave-assisted digestion also exploits high pressure applied in the closed vessels to accelerate digestion [10]. Alternatively, applying high temperature up to 600°C to the sample, or dry digestion, is suitable for samples that need pre-concentration. The risk of sample loss and longer digestion time can often be a hindrance to the application of the dry digestion method [11].

AAS, ICP-AES, and ICP-MS are the most common atomic spectroscopy instruments available. The instrumentation of atomic spectroscopy is generally divided into five main parts: sample introduction, excitation source, spectroscope/spectrometer, detector, and signal processing. The sample, usually in the form of solution, is pumped using a peristaltic pump into the nebulizer to form aerosol before entering the instrument. A solid sample can be analyzed using a graphite furnace-atomic absorption spectroscopy (GF-AAS). In recent years, laser ablation coupled to ICP-MS has become more popular for the analysis of solid samples due to much shorter preparation time and the possibility of producing an elemental map [12].

The excitation source in AAS depends on the type of instrument responsible for transforming the liquid (or solid) samples into gas, followed by atomizing and ionizing the samples. In flame-AAS, the excitation source is a flame, which is produced by burning air-acetylene gas or nitrous oxide-acetylene gas to reach 2300°C or 2700°C, respectively. Air-acetylene gas is commonly used, while nitrous oxide-acetylene gas is more suitable for samples that easily form oxides. GF-AAS can reach up to 3000°C, resulting in a much lower detection limit and a higher sensitivity compared to flame-AAS. ICP is even more sensitive because the temperature of the plasma can reach 10 000 K [13].

Another critical point to consider for choosing the instruments is the number of elements analyzed. For one or few elements analyzed in a sample, AAS is the most economical choice if the limit of detection permits. AAS is the cheapest instrument for metal analysis, compared to ICP-AES and ICP-MS. The light source of the AAS is usually a hollow cathode lamp that emits light at a specific wavelength, depending on the element analyzed. The wavelength (in multi-element hollow cathode lamp) or the lamp itself (in single-element hollow cathode lamp) should be changed once the element analyzed is changed. This step requires additional time if multiple elements are analyzed. Therefore, AAS is only suitable for analyzing one or few elements in a sample. To analyze multiple elements in many samples, ICP-based instruments are more efficient than AAS, although the purchasing and operational costs are high. In ICP-AES, the light emitted by the ions, following the excitation in the ICP, is directly analyzed by the detector. In ICP-MS, the ions are directly analyzed in a mass spectrometer. Thus, many elements can be directly measured in a sample consecutively using an ICP-AES or ICP-MS. In the context of fruit quality assurance, the majority of the elemental analysis in fruit utilizes ICP-AES to determine the concentration of several elements in each sample [14].

10.3 Quality assurance

A general consensus for the evaluation of fruit quality using elemental content has not been established. The pattern of elemental concentration in the fruit produced over a period of time or with different crop management is normally compared using statistical analysis. Alternatively, deviation from optimum index (DOP), introduced by Montañés *et al* [15], can be used to estimate the adequacy of nutrient content. The DOP (%) is calculated with the following equation:

$$\text{DOP} = \left(\frac{C \times 100}{C_{\text{ref}}}\right) \times 100$$

where C is the concentration of element of interest in the sample and C_{ref} is the optimum concentration of element of interest. Meanwhile, C and C_{ref} are the concentration in the dry matter, ideally determined with the same conditions [15]. The combination of several DOP values is expressed as ΣDOP. A negative DOP value indicates insufficient nutrient, while a positive DOP value suggests nutrient excess in the sample [4]. This method is simple and straightforward, and therefore it is normally used for routine interpretation. However, the determination of reference value could be tricky because the growing condition between the reference and the sample should be similar and enough data should be gathered.

Certain elements, such as heavy metals, have safe limitation values in fruit to avoid unwanted health effects. The monitoring of these elemental values in fruit is one of the quality control aspects that is sometimes overlooked, especially in traditional/local markets that do not carry out rigorous quality control procedures. For example, the fruits found in a market in North India contain a safe level of Cu (less than 40 $\mu g\ g^{-1}$) but not Cd and Cr (more than 0.2 and 2.3 $\mu g\ g^{-1}$, respectively) according to the recommended maximum allowable limit established by Food and Agriculture Organization (FAO)/World Health Organization (WHO). Meanwhile, Cd and lead (Pb) concentration in several fruits marketed in Poland were mostly below the maximum permissible concentration in fruits as determined with graphite furnace AAS. Only 1% of the samples were found to have higher concentration of Cd than the maximum permissible level [16]. Nevertheless, this data suggests that the levels of toxic elements should be routinely monitored to prevent health problems in the population.

10.4 Applications of elemental contents for quality assurance and authentication analysis of fruit juices

The multi-element composition of fruit juices is essential for the quality control of fruit juices. In addition, the elemental contents may be used as tools for the authentication of fruit juices and could be used as a powerful geographical indicator of the origin of the juice since the elemental compositions reflect the soil composition of the location where the fruits are grown [17]. Since a large number of variables are involved during applying elemental compositions for characterization of fruit juices, advanced chemometric analysis is essential to evaluate the information [18]. Table 10.1 compiled the application of elemental analysis based on atomic spectroscopy in fruits and fruit juices.

Table 10.1. Application of elemental analysis based on atomic spectroscopy in fruits and fruit juices.

Samples	Analytical technique	Preparation technique	Elements	References
Fresh olive fruits	Flame atomic absorption spectrometry	Wet-ashing	K, Na, Ca, Mg, Zn, Cu, Fe, and Mn	[26]
Commercial fruit juices (cherry, apple, orange, peach, and apricot)	Flame atomic absorption spectrometry	Dilution for Na, Mg, Ca, and K Direct measurement for Cu, FE, Cr, Mn, Zn, and Pb	Cu, Fe, Cr, Mn, Zn, Pb, Na, Mg, Ca, and K	[27]
Fruits (Citrus limetta, kiwi, and pomegranate)	Flame atomic absorption spectrometry	Dispersive solid phase extraction	Pb and Cd	[28]
Grape juices and wines	Flame atomic absorption spectrometry	Hot digestion with HNO_3 and H_2O_2	Cu, Fe, and Mn	[29]
Edible seeds (amaranth, chia, pumpkins, quinoa, and sunflower seeds)	High-resolution continuum source atomic absorption spectrometry	Hot digestion with HNO_3 and H_2O_2	Cd, Ni, and Fe	[30]
Banana, kiwi, mandarin	Flame atomic absorption spectrometry	Hot digestion with HNO_3 and H_2O_2	Cd, Cu, Zn, and Mn	[31]
Avocado fruit	Flame atomic absorption spectrometry	Digestion with nutric acid–sulfuric acid–perchloric acid mixture (3:1:1 v/v/v)	Na, K, Fe, Ca, and Mg	[32]
Sour cherry and table grape	ICP-OES	Dry ashing. The ashes obtained were dissolved in 5% (v/v) HNO_3	Na, K, Ca, Mg, Fe, Cu, Zn, Mn, Cr, Cd, Co, Pb, and Ni	[7]
Citrus fruits	ICP-OES and ICP-MS	Microwave digestion	K, Ca, Zn, Mn, Pb, Cd, Al, and As	[33]
Tamarind, star fruit, golden berry, kumquat, dragon fruit, and passion fruit	ICP-OES	Microwave acid digestion	Al, B, Ca, Cu, Fe, K, Mg, Mn, Na, Zn, P, and S	[34]

(*Continued*)

Table 10.1. (*Continued*)

Samples	Analytical technique	Preparation technique	Elements	References
Fruits juices and fruits canned (Peach, orange, cherry, and pineapple)	ICP-OES	Microwave digestion	Cd, Hg, As, Sn, Pb, and Al	[35]
Strawberry and khaki	ICP-MS	Microwave acid digestion	Na, Mg, Al, P, S, K, Ca, V, Cr, Mn, Fe, Co, Cu, Zn, As, Se, Rb, Sr, Mo, Cd, Sn, Sb, Cs, Ba, Hg, Pb, and U	[36]
Fruit juices (orange, tangerine, red berries, peach, pineapple, apple, and grape, mango, and orange)	ICP-OES and ICP-MS	Microwave acid digestion	Al, As, Ca, Cd, Co, Cr, Cu, Fe, Mg, Mn, Na, Ni, Pb, V, Sb, Mo, and Zn	[37]
Saffron (*Crocus sativus*)	ICP-OES	Dry ashing. The ashes obtained were dissolved in HNO_3	Zn, Cu, Fe, K, Mn, Mg, Ca, Na, and P	[38]
Opuntia ficus-indica L. Miller fruits	ICP-MS	Microwave acid digestion	Na, K, Ca, Mg, Fe, Zn, Mn, As, Cu, Ni, Se, Cd, and Pb	[39]
Pomegranate juice	ICP-OES	Dilution in 0.2% HNO_3 (1:1, v/v), centrifugation (20 min, 2000 rpm)	Ca, Cd, Cu, Fe, K, Mg, Mn, Na, Ni, P, Pb, Sn, Zn	[40]
Apple, melon, grapefruit, mosambi, lime fruit, and guava	Flame atomic absorption spectrometry	Digestion using 30% H_2O_2 and 50% HNO_3	Cd, Zn, Ni, Co, Cu, Cr, and Mn	[41]
Apple, grape, raspberry, strawberry, pear, and cranberry	Flame atomic absorption spectrometry	Digestion using 30% H_2O_2 and 65% HNO_3	Cd and Pb	[16]
Mango, melon, and banana	Flame atomic absorption spectrometry	Dry ashing	Pb, Cu, Zn, Co, Ni, and Cd	[42]
Water melon, peach, banana, mango, and apple	Flame atomic absorption spectrometry	Digestion using nitric acid, perchloric acid and sulfuric acid with ratio of 5:1:1	Fe, Pb, Cd, Co, Cu, Ni, and Mn	[43]

10.4.1 Apple and apple juice

It has long been known that calcium (Ca) deficiency in apple results in a bitter pit, which is a common post-harvest disorder that is marked by brownish spots that are usually found on the surface of the fruit. Ca is involved in maintaining the plasma membrane, whose deficiency results in membrane leakage or plasmolysis, leading to cell death and necrotic tissue. Overall, Ca concentration in fruits with bitter pit could be half of those without bitter pit [19]. Imbalanced Ca distribution inside the fruit, evidenced by accumulation of Ca in vacuole and cell wall, also contributes to the development of bitter pit. The heightened Ca transport is reflected by the increased expression of vacuolar pyrophosphatase (PPase) and putative Ca^{2+}/H^{+} exchangers (CAX), which supports Ca influx to the vacuole, and putative pectin methyl esterases (PME), which increases Ca binding sites in the cell wall [20].

Ca deficiency also causes imbalance of other elemental concentrations. In conjunction with low Ca concentration in fruit, the potassium (K) and magnesium (Mg) concentration was significantly higher in apples with bitter pit than those without bitter pit, as analyzed with ICP-AES [19]. In addition to bitter pit, a high K/Ca ratio has been associated to post-harvest disorders and reduced storability [21]. K and Mg may interfere with Ca during absorption and transportation to other tissues. In the soil, high K and Mg levels hinder Ca absorption due to electrochemical imbalances [22]. The Ca intake into the apple fruits is also suboptimal in the presence of high concentration of K and Mg. Ca competes with the latter elements during the cellular intake but Ca role for structural support is irreplaceable.

Knowing that elemental imbalances in apple fruit could lead to lower quality of fruit, the elemental content in the tree often needs to be monitored during the pre-harvest period so that the nutritional deficiency can be improved. The indication of elemental adequacy in apples can be reflected from the elemental concentration in the leaves. The P, K, and Mg levels, but not N and Ca levels, in apple leaves vary according to the fertilizer treatment. Nevertheless, fertilizer treatment was not adequate to increase the nutrient status of the tree, which is reflected by negative DOP values of N, P, K, and Ca [23].

The combination of isotope ratios of $\delta^{13}C$, $\delta^{18}O$ along with the elemental contents of 32 elements has been successfully applied for the differentiation of apple fruits from different origins. In this study, the isotope rations and 28 elements were used as variable during the classification of apple fruits. Principal component analysis (PCA), partial least squared-discriminant analysis (PLS-DA), and OPLS-DA have been used for the classification of apple samples according to its origins. Among these variables, 10 variables—namely Tl, Se, δ18O, B, Mg, Sr, Nd, Mo, As, and Na—indicated the major variables contributing to the classification of the samples. These findings show that the stable isotopic ratios and the element compositions were able to differentiate apple samples according to geographic origins [24]. The combination of isotope ratios and elemental compositions is also successful for the verification of Chinese concentrated apple juice according to geographical regions.

Using the pattern recognition chemometrics of PCA and PLS-DA, the predictive models for classification of apple juices were fruitfully established with an accuracy rate of 93.2% [25].

10.4.2 Avocado and avocado fruit

Since elemental concentrations, especially Ca, in apple have been successfully established as one of the factors that contribute to the fruit quality, the correlation between elemental concentrations and fruit quality has been examined in other fruits. In avocado trees, Ca is distributed in all parts of the tree, with leaves, branches, and roots having the highest Ca concentration, while Ca concentration in fruit and wood are among the lowest. K has the same pattern as Ca, but Mg is the opposite. In the fruit, the Ca concentration in the pulp is higher than that in the peel, but the K concentration in the peel is higher. The absolute concentrations of the elements vary according to the cultivar of the avocado [44].

Ca concentration in avocado determines the post-harvest quality of the fruit. High Ca concentration is associated with prolonged storage life of avocado, due to delayed ripening process as the ethylene production and respiratory rates were reduced. The post-harvest disease incidence was found to be low in avocado with high Ca and Ca/K ratio. Indeed, deviated Ca level in the fruit causes disturbance to the transport of other elements. The low Ca concentration in epidermis has been correlated with chilling injury, pericarp damage, and irregular ripening [45]. Although irregular ripening can happen when the dry matter in the fruit is too low or too high, low Ca concentration in avocado is known to cause lower dry matter content (<23%) [46].

Micronutrients also play a vital role in the tree's growth, which eventually affects avocado quality. Boron (B) is known to affect pollination during fruit production. B increases the yield of avocado and produces heavier fruit in which the mass of the flesh, but not the seed, increases. Moreover, the ratio of unsaturated to saturated fatty acid increases, so that the fruit provides superior health benefits [47]. Macronutrients and micronutrients, except for Fe and manganese (Mn), are accumulated in the fruit with the highest concentration in the peel and pulp. The concentration of nutrients in fruit is dynamic and can be affected by environmental conditions, such as rainfall and other climatic characteristics, water availability in the soil, and fertilizer treatment. Typically, the nutrient levels decrease during the year due to these environmental factors [44].

The stable isotope composition of five elements (C, N, S, H, and O) and the elemental compositions have been successfully applied as analytical tools for the authentication of avocado. In this study, the authors collected 131 avocados from different producing regions over three years to study the main geographical differences. Classification of avocado fruits was carried out using PLS-DA model. The results proved that Spanish and non-Spanish avocados can be distinguished using a combination of stable isotopes and elemental profiles with 98% correct classification, indicating high prediction accuracy. Eight elements—namely Fe, Ca, Mn, Sr, Rb,

Eu, Ba, and Mg—offered the most elements contributing to the classification of avocado fruits from different regions [48].

10.4.3 Other fruits

Soil condition heavily influences fruit elemental concentrations. Increased irrigation, such as in high rainfall, reduces Ca absorption in tomatoes, causing blossom-end rot disease in tomato plants. The blossom end of the fruit has been found to have the lowest Ca concentration, compared to the stem end of the fruit [49]. Calcareous soil is known to have low Mn concentration, resulting in hampered nectarine quality and yield. Furthermore, the high concentration of Ca in the soil hindered Mn uptake. Besides low Mn, calcareous soil causes low elemental solubility, thus reducing the absorption of various elements into the nectarine plants, as determined with ICP-AES and AAS in the leaves. The ΣDOP index shows that the elemental contents in the plant grown in calcareous soils were imbalanced [50].

In addition to inherent soil composition, beneficial microorganisms in the soil can affect fruit production. In tomato plants, the arbuscular mycorrhizal fungi, inoculated in the soil during seeding, altered the plant's elemental condition. The concentration of K, Ca, Mg, P, and sulfur (S) in the fruits were higher than those non-inoculated control, while Na concentration was lower than the control. The soil microbes improve the solubility and availability of minerals, so that more minerals can be transported into the fruit. The overall quality of the fruit was improved, marked by significantly increased dry residue, soluble solids, and organic acids in fruit inoculated with beneficial microorganisms [51].

The cultivar of the plant, shade intensity, and fruit bagging also affect the elemental concentration in the plant. Reig *et al* [4] used elemental concentration in a plant's leaves to estimate the quality of plums. The N concentration can indicate the fruit quality because the N level inversely correlates with soluble solids amount. The concentrations of P and N were determined by the cultivar, but not the rootstock of the plant. The concentrations of P and N were also influenced by shade intensity. Pineapple plants with lower light intensity have an increased P and N uptake. The K uptake, however, was optimum with moderate light intensity. Nevertheless, the highest pineapple yield was produced in the plants with moderate (50%) shade, in which biomass was optimally produced [52]. Fruit bagging, which is used to protect fruit from external damage and diseases, influences Ca concentration in mango. Although the Ca concentration was lower in bagged mango, the quality of the fruit was generally not affected [53].

The elemental compositions as determined using AAS, AES, ICP-MS, and stable isotope ratio present in fruit are unique; therefore, these compositions could be used as an analytical tools for the authentication of fruit juices, especially for the geographical origin designation. Twenty-five elements of Ag, Al, As, Ba, Bi, Co, Cr, Cu, Fe, Ga, In, La, Li, Mn, Mo, Ni, Rb, Sb, Sc, Se, Sn, Sr, Tl, V, and Zn have been determined using ICP-MS, and the levels of element contents have been used as variables for the differentiation and geographical origins of lemon juice samples from three different provinces of Argentina. The chemometrics of supervised pattern recognition, including linear discriminant analysis (LDA), partial least square discriminant analysis (PLS-DA), k-nearest neighbors (k-NN), random forest (RF), and support vector machine with radial basis function Kernel (SVM), have been used and optimized to select the best classification models. Among these supervised pattern recognition techniques, SVM provided the better classification performance with correct classification rate and sensitivity rate of 76.2% and specificity of 87.5%, compared to RF, k-NN, LDA, and PLS-DA, listed in descending order. These results confirm that the combination of trace element contents with SVM technique offered valuable tools to evaluate the geographical origin of lemon juice samples [54].

Stable isotope ratios based on isotopes of $\delta^{13}C$, $\delta^{2}H$ and $\delta^{18}O$ combined with elemental analysis have been successfully applied to determine the geographical origin of peach samples. During this study, 150 peach samples from six regions in China was analyzed, and the contents of 18 elements were subjected to analysis using ICP-MS to develop classification models using PCA, stepwise linear discriminant analysis (SLDA), and orthogonal partial least square discriminant analysis (OPLS-DA) for the differentiation and classification of peach fruits. The results showed that PCA can differentiate the fruit samples cultivated from coastal and inland regions, mainly due to H and O isotope ratios. SLDA and OPLS-DA provided a clear discrimination of the peach samples from different regions. The important variables responsible for the differentiation in these models included $\delta^{2}H$, $\delta^{13}C$ and some elements (Na, Cd, V, Al, $\delta^{18}O$, Be, K, Co, Cu, and Ag). These findings indicated that isotope ratios and elemental analysis combined with suitable chemometrics can be a powerful analytical technique to determine the geographical origins of peach fruit juices [8].

10.5 Conclusion

Elemental adequacy determines fruit yield and quality. A deficiency of a certain element may cause disease, e.g., Ca deficiency causes bitter pit in mango. Imbalances of elemental contents in fruit can also happen when a certain element dominates the uptake. The determination of the elemental levels in fruit is mostly performed using ICP-AES because several elements are usually assessed, although AAS and ICP-MS are also used. Although elemental levels are critical in determining fruit quality, standardized criteria to determine elemental adequacy have not yet been established. Therefore, a comprehensive guideline on elemental adequacy in fruits to assess fruit quality would be practical for routine analysis.

References

[1] Jivan C and Sala F 2014 Relationship between tree nutritional status and apple quality *Hortic Sci.* **41** 1–9

[2] Musacchi S and Serra S 2018 Apple fruit quality: overview on pre-harvest factors *Sci. Hortic (Amsterdam).* **234** 409–30

[3] Verma Y and Rana S V S 2014 Assessment of cadmium, chromium, and copper levels in market fruit samples in Meerut, North India *Toxicol. Environ. Chem.* **96** 1516–22

[4] Reig G, Font i Forcada C, Mestre L, Jiménez S, Betrán J A and Moreno M Á 2018 Horticultural, leaf mineral and fruit quality traits of two 'Greengage' plum cultivars budded on plum based rootstocks in Mediterranean conditions *Sci. Hortic (Amsterdam).* **232** 84–91
[5] Preti R 2019 Progress in beverages authentication by the application of analytical techniques and chemometrics *Quality Control in the Beverage Industry: Volume 17: The Science of Beverages* (Amsterdam: Elsevier) pp 85–121
[6] Tan M, Sudjadi , Astuti and Rohman A 2018 Validation and quantitative analysis of cadmium, chromium, copper, nickel, and lead in snake fruit by inductively coupled plasma-atomic emission spectroscopy *J. Appl. Pharm. Sci.* **8**
[7] Mitić S S, Obradović M V, Mitić M N, Kostić D A, Pavlović A N and Tošić S B *et al* 2012 Elemental composition of various sour cherry and table grape cultivars using inductively coupled plasma atomic emission spectrometry method (ICP-OES) *Food Anal. Methods* **5** 279–86
[8] Li A, Zhao J, Xi J, Yang X, Jin X and Chen Q *et al* 2021 Geographical authentication of peach in China based on stable isotope combined with multielement analysis of peach juice *Food Control.* **127**
[9] Aceto M 2016 The use of ICP-MS in food traceability *Advances in Food Traceability Techniques and Technologies: Improving Quality Throughout the Food Chain* (Amsterdam: Elsevier) pp 137–64
[10] Gholami M, Behkami S, Zain S M and Bakirdere S 2016 A simple design for microwave assisted digestion vessel with low reagent consumption suitable for food and environmental samples *Sci. Rep.* 6–8
[11] Thirumdas R, Janve M, Siliveru K and Kothakota A 2019 Determination of food quality using atomic emission spectroscopy *Evaluation Technologies for Food Quality* (Amsterdam: Elsevier) pp 175–92
[12] Costas-Rodríguez M, Van Acker T, Hastuti A A M B, Devisscher L, Van Campenhout S and Van Vlierberghe H *et al* 2017 Laser ablation-inductively coupled plasma-mass spectrometry for quantitative mapping of the copper distribution in liver tissue sections from mice with liver disease induced by common bile duct ligation *J. Anal. At. Spectrom.* **32** 1805–12
[13] Soodan R K, Pakade Y B, Nagpal A and Katnoria J K 2014 Analytical techniques for estimation of heavy metals in soil ecosystem: a tabulated review *Talanta* **125** 405–10
[14] Huang J, Hu X, Zhang J, Li K, Yan Y and Xu X 2006 The application of inductively coupled plasma mass spectrometry in pharmaceutical and biomedical analysis *J. Pharm. Biomed. Anal.* **40** 227–34
[15] Moñtanés L, Heras L, Abadia J and Sanz M 1993 Plant analysis interpretation based on a new index: deviation from optimum percentage (DOP) *J. Plant Nutr.* **16** 1289–308
[16] Rusin M, Domagalska J, Rogala D, Razzaghi M and Szymala I 2021 Concentration of cadmium and lead in vegetables and fruits *Sci. Rep.* **11** 1–10
[17] Dasenaki M E and Thomaidis N S 2019 Quality and authenticity control of fruit juices-a review *Molecules* **24**
[18] Drivelos S A and Georgiou C A 2012 Multi-element and multi-isotope-ratio analysis to determine the geographical origin of foods in the European Union *TrAC—Trends Anal. Chem.* **40** 38–51
[19] Bonomelli C, Mogollón R, de Freitas S T, Zoffoli J P and Contreras C 2020 Nutritional relationships in bitter pit-affected fruit and the feasibility of vis-NIR models to determine calcium concentration in 'Fuji' Apples *Agronomy* **10**

[20] de Freitas S T, Amarante C V T D, Labavitch J M and Mitcham E J 2010 Cellular approach to understand bitter pit development in apple fruit *Postharvest Biol. Technol.* **57** 6–13
[21] Samnegård U, Hambäck P A and Smith H G 2019 Pollination treatment affects fruit set and modifies marketable and storable fruit quality of commercial apples *R. Soc. Open Sci.* **6**
[22] Zharare G E, Asher C J and Blamey F P C 2011 Magnesium antagonizes Pod-Zone calcium and zinc uptake by developing peanut pods *J. Plant Nutr.* **34** 1–11
[23] Milošević T and Milošević N 2015 Apple fruit quality, yield and leaf macronutrients content as affected by fertilizer treatment *J. Soil Sci. Plant Nutr.* **15** 76–83
[24] Liu X, Mu J, Tan D, Mao K, Zhang J and Ahmed Sadiq F *et al* 2022 Application of stable isotopic and mineral elemental fingerprints in identifying the geographical origin of concentrated apple juice in China *Food Chem.* **391** 133269
[25] Liu X, Zhao Y, Qi P, Liu Y, Li X and Deng W *et al* 2022 Origin verification of Chinese concentrated apple juice using stable isotopic and mineral elemental fingerprints coupled with chemometrics *J. Food Compos. Anal.* **109** 104424
[26] Fernandez-Hernandez A, Mateos R, Garcia-Mesa J A, Beltran G and Fernandez-Escobar R 2010 Determinación del contenido de elementos minerales en aceituna mediante espectrofotometría de absorción atómica *Spanish J Agric. Res.* **8** 1183–90
[27] Ince H and Coskun N 2008 Determination of heavy metals in fruit juices by flame atomic absorption spectrometry *Asian J. Chem.* **20** 3537–42
[28] Behbahani M, Ghareh Hassanlou P, Amini M M, Omidi F, Esrafili A and Farzadkia M *et al* 2015 Application of solvent-assisted dispersive solid phase extraction as a new, fast, simple and reliable preconcentration and trace detection of lead and cadmium ions in fruit and water samples *Food Chem.* **187** 82–8
[29] Dutra M C P, Rodrigues L L, de Oliveira D, Pereira G E and Lima M S 2018 Integrated analyses of phenolic compounds and minerals of Brazilian organic and conventional grape juices and wines: validation of a method for determination of Cu, Fe and Mn *Food Chem.* **269** 157–65
[30] dos Santos N K V, dos Santos L, Damin I C F, Vale M G R and Dessuy M B 2022 Multielement determination of metals in edible seeds by HR-CS GF AAS and direct analysis *J. Food Compos. Anal.* **111**
[31] Soceanu A 2010 Determination of some trace metal concentrations in imported fruits by F-AAS and ICP-MS *Environ. Eng. Manag. J.* **9** 1039–44
[32] Maitera O, Osemeahon S A and Barnabas H L Proximate and elemental analysis of avocado fruit obtained from Taraba State, Nigeria
[33] Hong Y S, Choi J Y, Nho E Y, Hwang I M, Khan N and Jamila N *et al* 2019 Determination of macro, micro and trace elements in citrus fruits by inductively coupled plasma–optical emission spectrometry (ICP-OES), ICP–mass spectrometry and direct mercury analyzer *J. Sci. Food Agric.* **99** 1870–9
[34] Karasakal A 2021 Determination of major, minor, and toxic elements in tropical fruits by ICP-OES after different microwave acid digestion methods *Food Anal. Methods* **14** 344–60
[35] Fathabad A E, Shariatifar N, Moazzen M, Nazmara S, Fakhri Y and Alimohammadi M *et al* 2018 Determination of heavy metal content of processed fruit products from Tehran's market using ICP- OES: a risk assessment study *Food Chem. Toxicol.* **115** 436–46
[36] Potočnik D, Jagodic Hudobivnik M, Mazej D and Ogrinc N 2021 Optimization of the sample preparation method for determination of multi-elemental composition in fruit samples by ICP-MS analysis *Meas. Sens.* **18** 100292

[37] de Souza M J B, Barciela-Alonso M C, Aboal-Somoza M and Bermejo-Barrera P 2021 Determination of the trace element contents of fruit juice samples by ICP OES and ICP-MS *Brazilian J Anal Chem.* **9** 49–61

[38] Ibourki M, Gharby S, Sakar E H, Hani O E, Digua K and Amine A *et al* 2022 Elemental profiling and geographical differentiation of saffron (*Crocus sativus* L.) using inductively coupled plasma-optical emission spectroscopy (ICP-OES) and principal component analysis *Chem Data Collect* **41** 100937

[39] Mottese A F, Naccari C, Vadalà R, Bua G D, Bartolomeo G and Rando R *et al* 2018 Traceability of *Opuntia ficus-indica* L. Miller by ICP-MS multi-element profile and chemometric approach *J. Sci. Food Agric.* **98** 198–204

[40] Akpinar-Bayizit A 2010 Analysis of mineral content in pomegranate juice by ICP-OES *Asian J. Chem.* **22** 6542–6

[41] Yaqub G, Khan A, Zishan Ahmad M and Irshad U 2021 Determination of concentration of heavy metals in fruits, vegetables, groundwater, and soil samples of the cement industry and nearby communities and assessment of associated health risks *J. Food Qual.* **2021**

[42] Elbagermi M A, Edwards H G M and Alajtal A I 2012 Monitoring of heavy metal content in fruits and vegetables collected from production and market sites in the Misurata Area of Libya *ISRN Anal. Chem.* **2012** 1–5

[43] Un Nisa K, Samiullah S, Khan N and ur-Rehman A 2020 Detection of heavy metals in fruits and vegetables available in the market of Quetta City *Al-Nahrain J. Sci.* **23** 47–56

[44] Aguirre-Arcos A, García-Carmona M, Reyes-Martín M P, San-Emeterio L M, Fernández-Ondoño E and Ortiz-Bernad I 2022 Effects of Pruning Mulch on nutrient concentration of avocado (*Persea americana* Mill.) fruit under subtropical conditions *Horticulturae* **8** 1–11

[45] Kämper W, Trueman S J, Tahmasbian I and Bai S H 2020 Rapid determination of nutrient concentrations in hass avocado fruit by vis/NIR hyperspectral imaging of flesh or skin *Remote Sens.* **12** 1–19

[46] Ramírez-Gil J G, Henao-Rojas J C and Morales-Osorio J G 2021 Postharvest diseases and disorders in avocado cv. Hass and their relationship to preharvest management practices *Heliyon* **7** e05905

[47] Hapuarachchi N S, Kämper W, Wallace H M, Hosseini Bai S, Ogbourne S M and Nichols J *et al* 2022 Boron effects on fruit set, yield, quality and paternity of hass avocado *Agronomy* **12** 1–15

[48] Muñoz-Redondo J M, Bertoldi D, Tonon A, Ziller L, Camin F and Moreno-Rojas J M 2022 Multi-element and stable isotopes characterization of commercial avocado fruit (*Persea americana* Mill) with origin authentication purposes *Food Control.* **137**

[49] Taylor M D, Locascio S J and Alligood M R 2004 Blossom-end rot incidence of tomato as affected by irrigation quantity, calcium source, and reduced potassium *HortScience* **39** 1110–5

[50] Mestre L, Reig G, Betrán J A, Pinochet J and Moreno M Á 2015 Influence of peach-almond hybrids and plum-based rootstocks on mineral nutrition and yield characteristics of 'Big Top' nectarine in replant and heavy-calcareous soil conditions *Sci. Hortic. (Amsterdam)* **192** 475–81

[51] Sellitto V M, Golubkina N A, Pietrantonio L, Cozzolino E, Cuciniello A and Cenvinzo V *et al* 2019 Tomato yield, quality, mineral composition and antioxidants as affected by beneficial microorganisms under soil salinity induced by balanced nutrient solutions *Agriculture* **9** 1–15

[52] Kishore K, Rupa T R and Samant D 2021 Influence of shade intensity on growth, biomass allocation, yield and quality of pineapple in mango-based intercropping system *Sci. Hortic. (Amsterdam)* **278** 109868

[53] Hofman P J, Smith L G, Joyce D C, Johnson G I and Meiburg G F 1997 Bagging of mango (*Mangifera indica* cv. 'Keitt') fruit influences fruit quality and mineral composition *Postharvest Biol. Technol.* **12** 83–91

[54] Gaiad J E, Hidalgo M J, Villafañe R N, Marchevsky E J and Pellerano R G 2016 Tracing the geographical origin of Argentinean lemon juices based on trace element profiles using advanced chemometric techniques *Microchem. J.* **129** 243–8

IOP Publishing

Advances in Spectroscopic Analysis of Food and Drink

Ashutosh Kumar Shukla

Chapter 11

Spectroscopic techniques employed for quality assessment of edible oils

Sumaiya Fatima and Vivek Kumar

Edible oils are subjected to adulteration and misbranding, which is important nutritionally and economically. Therefore, authentication, quality assessment, and adulteration detection of vegetable oils is a primary concern for consumers, manufacturers, and traders. As a result, numerous analytical techniques, including standard chemical test, chromatography, spectroscopy, and so on, have been developed to determine the quality and authenticity of edible oils. Meanwhile, spectroscopic methods provide a rapid, dependable, non-destructive, cost-effective, and real-time measuring tool to monitor and assess oil quality. Some of the spectroscopic techniques used in edible oil verification include UV–visible (UV–Vis) spectroscopy, infrared (IR) spectroscopy, Fourier transform infrared (FTIR) spectroscopy, Raman spectroscopy, nuclear magnetic resonance (NMR) spectroscopy, and fluorescence spectroscopy. These techniques can be used to estimate the physiochemical characteristics of oil, detect its bioactive components, determine its geographical origin and cultivar, and detect adulteration and lipid oxidation in diverse oil samples. This chapter will discuss the various spectroscopic techniques and their applications in determining the quality of edible oils.

11.1 Introduction

Vegetable oils are important lipid sources in the human diet, and have substantial nutritional and calorific properties. These are vital macronutrients which act as an energy-reserve, insulate and protect vital organs, help in the transportation of fat-soluble vitamins, and have many other physiological functions [1]. Edible oil is important nutritionally, and its consumption and utilization are considerable worldwide. Thus, detecting its authenticity, adulteration, and quality is critical to human health [2]. However, due to their comparable appearance, color, and chemical characteristics, there is always a temptation to mislabel or adulterate expensive oils

doi:10.1088/978-0-7503-5573-5ch11 © IOP Publishing Ltd 2023

with similar cheaper edible oils for the sake of profit. Despite widespread public awareness of the scale and risks of oil adulteration, the frequency of deception is difficult to measure [3]. Therefore, various analytical approaches to assess the authenticity and quality of edible oils have been developed. These methods encompass chemical estimation, chromatographic techniques, use of electronic nose and tongue, immunological investigations, and other techniques [4–6]. However, most chemical and chromatographic procedures have limitations, e.g., they are destructive, time-consuming, and complex. Consequently, a rapid, reliable, non-destructive, cost-effective, and real-time measuring method is required [7]. Spectroscopic techniques have emerged as sustainable analytical tools for the precise real-time assessment of the quality of edible oil. Among the spectroscopic techniques used in edible oil authentication are ultraviolet–visible (UV–Vis), infrared (IR), Fourier transform infrared (FTIR), Raman, NMR, and fluorescence spectroscopy [8]. Therefore, this chapter will discuss the various spectroscopic techniques and their application in assessing the quality of edible oils.

11.2 Spectroscopic techniques

Spectroscopy is concerned with the generation, measurement, and interpretation of spectra resulting from the test sample's interactions with electromagnetic radiation. UV–Vis, IR, Raman, NMR and fluorescence spectroscopy are all common spectrochemical analysis techniques. These spectroscopic approaches are frequently integrated with chemometrics, which is a chemical science that analyzes massive amounts of data, and establishes connections using mathematical and statistical tools. Spectroscopy, in general, is employed by the analyst to measure the amount of radiation absorbed or emitted by the analyte. This approach relies on the notion that matter's energy content is quantized and that it can absorb or emit photons of radiation if the energy associated with the photon equals the energy difference for permissible transitions of that particular species. These approaches differ in terms of the radiation wavelengths (figure 11.1) employed in the analysis and the molecular vs. atomic nature of the analyte [9].

11.2.1 UV–Vis spectroscopy

UV–Vis spectroscopy is an electronic spectroscopy that uses photons with wavelengths in the ultraviolet (100–400 nm) and visible (400–800 nm) ranges. It is based on Beer–Lambert's law, which states that when UV or visible light falls on sample, it absorbs it, and this absorbance depends on the amount of its constituent absorbing species. The absorbance of light induces change in the electronic state of molecules because the electrons absorb energy and move to a higher energy state from ground state orbital [10]. The fundamental concept of a spectrophotometer is that UV–visible light travels through a solvent and then lands on a photoelectric cell, which transforms radiant energy to electrical energy and is recorded by a galvanometer. The observed changes are fed to a data processing system and the results are then displayed in form of absorbance spectrum (figure 11.2) [11].

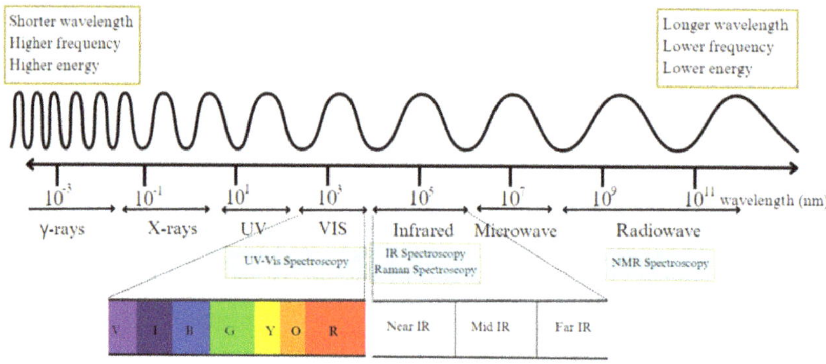

Figure 11.1. Overview of various spectroscopic techniques used for food analysis.

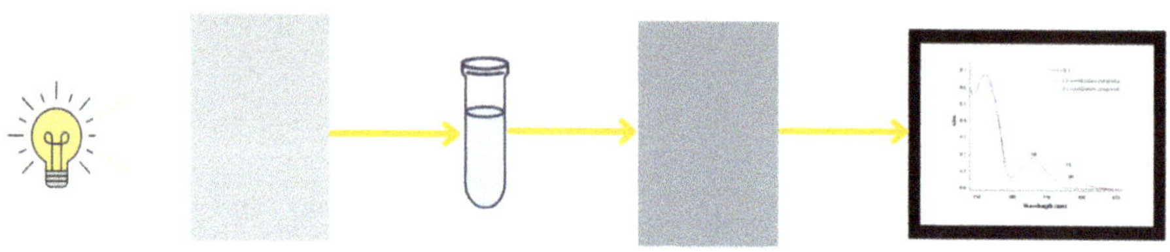

Figure 11.2. Schematic diagram to show the working of UV spectroscopy.

11.2.2 IR spectroscopy

IR spectroscopy, also known as vibrational spectroscopy, is an analytical technique centered around the interaction of a sample with the IR rays, followed by the detection and spectrum analysis of transmission or reflection radiation [12]. The sample absorbs IR rays via molecular vibrations divided by energies linked with the infrared component of electromagnetic radiation. Radiation with a particular frequency specifies a distinct chemical bond, and thus provides information related to the functional groups present in the molecule, which in turn are responsible for different chemical properties of the samples [13, 14]. IR radiation is divided into three types: near-infrared (NIR), mid-infrared (MIR), and far-infrared (FIR), these regions of the electromagnetic spectrum have wavelengths ranging from 0.8 to 2.5 μm (14 000–4000 cm^{-1}), 2.5–25 μm (4000–400 cm^{-1}), and 25–100 μm (400–10 cm^{-1}), respectively (figure 11.3). NIR and MIR are most typically employed in food analysis because of the absorption of organic compounds. Meanwhile, lower energy far-IR is often absorbed by heavy molecules such as organometallic and inorganic compounds [15], restricting their use in food analysis. Moreover, the MIR region gives more chemical information than the NIR region because the basic vibrations are detected in the MIR rather than the mixture of fundamental vibrations and complex overtones being received in the NIR region. Thus, the spectral peaks in the MIR are sharper and more defined than in the NIR spectra. Nowadays, practically all IR spectrometers used for analysis are clubbed with Fourier

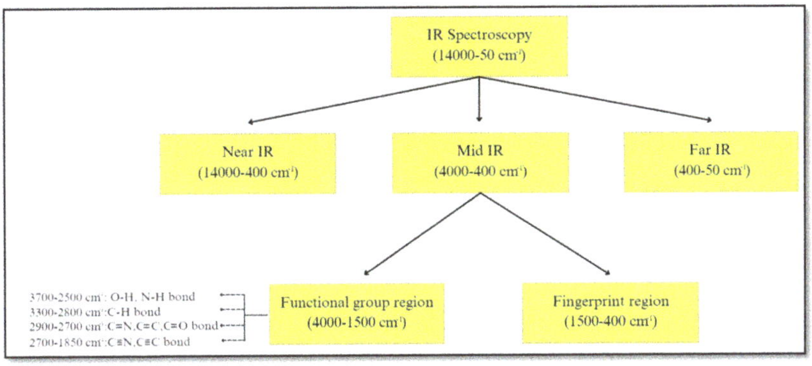

Figure 11.3. IR regions of the electromagnetic spectrum.

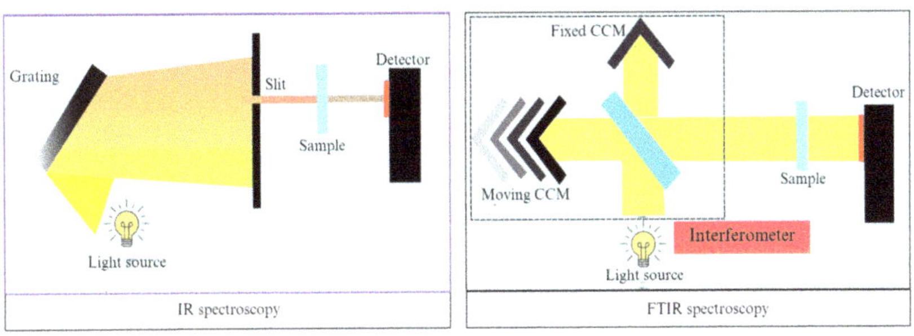

Figure 11.4. A comparison between IR and FTIR spectroscopy.

transform (FT-IR) rather than dispersive IR, owing to its multiple significant benefits over dispersive IR spectroscopy. FTIR spectroscopy uses an interferometer instead of a prism or grating, and thus generates a high-quality spectrum; has better data collection, speed, repeatability, and maintenance; and finds better applications (figure 11.4) [16].

11.2.3 Raman spectroscopy

Raman spectroscopy is based on the Raman scattering effect. It is primarily based on the rate of change of molecular polarizability or electron distribution composition of a molecule as oscillatory transitions that occur in the electronic ground state within the molecule [17]. Hence, Raman spectroscopy is based on measuring the shift in the energy of an outgoing photon that is produced when a photon from incoming light strikes a molecule (figure 11.5). The dispersed light carries information of the vibrational band energies of the molecules, which is affected by the chemical composition of the molecules [18]. Raman spectroscopy is thus widely used in food analysis to deliver a structural trace that can be used to identify compounds. Furthermore, when used in conjunction with Fourier transform (FT-Raman spectroscopy), it gives better information on molecular vibrations because adding Fourier transform is beneficial to remove background fluorescence [19].

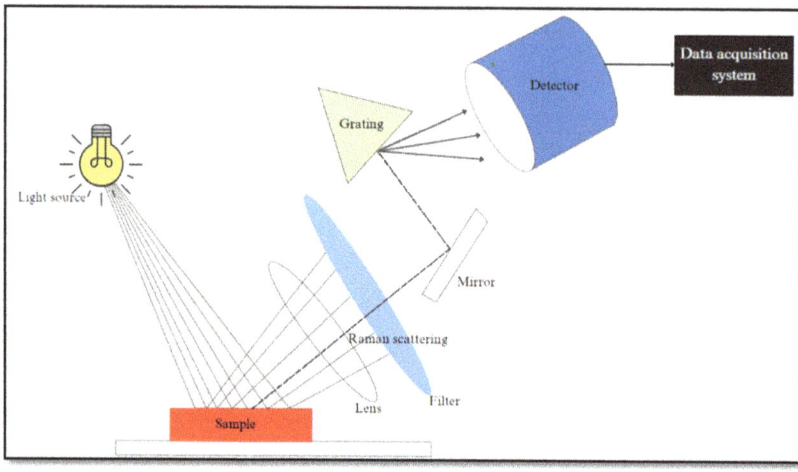

Figure 11.5. Diagrammatic representation of Raman spectroscopy.

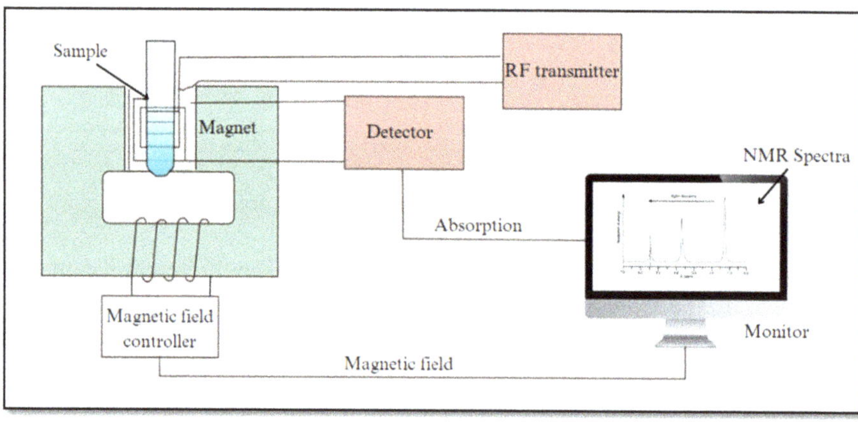

Figure 11.6. Schematic diagram of NMR spectroscopy.

11.2.4 NMR spectroscopy

NMR spectroscopy is a key analytical method that is used in organic chemistry to analyze the structure of molecules, as well as their interactions, kinetics, and dynamics [20]. The fundamental principle of spectroscopy is that the chemical composition and structure of various substances can be identified by their nuclei under the effect of an applied magnetic field. This spectroscopic technique uses a magnetic field along with some special detectors to evaluate changes by analyzing them [21]. The applied magnetic field forces the charged nucleus to move from a lower energy level to a higher energy level. The magnitude of the magnetic field and moment of the nuclear field determine the difference between the energy levels (figure 11.6). The electromagnetic radiation cycle generates the NMR signal caused by migration of nuclei to a higher energy level; however, the nuclei relax and achieve

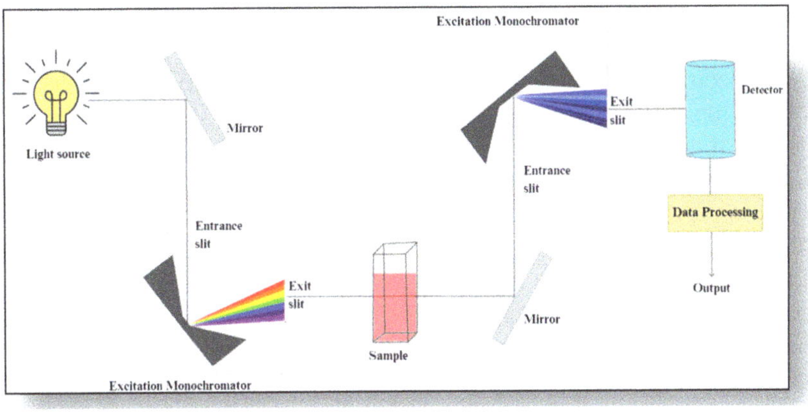

Figure 11.7. Schematic diagram of fluorescence spectroscopy.

thermal equilibrium when radiation is ceased. This energy released from the nuclei is recorded and fed to a computer to generate a spectrum. The most prevalent types of NMR are proton ^{1}H, phosphorus ^{31}P, fluoride ^{19}F, and carbon ^{13}C spectroscopy [22].

11.2.5 Fluorescence spectroscopy

Fluorescence spectroscopy is an analytical method that examines the fluorescence of a sample produced by light-matter interaction [23]. Fluorescence is the release of light following the absorption of UV or visible light by a luminous molecule, known as a fluorophore. As a result, the fluorophore absorbs radiation energy, and thereafter releases it in the form of light emission at a higher wavelength. Thus, absorption and fluorescence emission occur over a range of wavelengths correlating to many vibrational transitions as components of a single electronic transition. Thus, excitation and emission spectra are acquired and analyzed to characterize the exact fluorescence properties of molecules (figure 11.7) [24].

11.3 Application of spectroscopic techniques in the quality assessment of edible oils

11.3.1 Estimation of physiochemical properties

Oil quality must be monitored and maintained during refining and storing to ensure quality, usefulness, and for economic reasons. However, there is no single recognized standard for determining the quality of edible oils because oils differ significantly in their physical, chemical, and nutritional properties based on the variety of oil. Nonetheless, in the edible oil business, FFA (free fatty acid) composition, PV (peroxide value: depicting reaction of primary oxidation products), iodine value (depicting: degree of unsaturation), p-anisidine value, and specific gravity (depicting: purity) are commonly utilized as significant quality indicators [25]. The physiochemical properties of oil can be assessed by spectroscopic techniques and with the use of chemometrics to design a prediction model for the aforementioned properties of oils (table 11.1).

UV–Vis spectroscopy has been used to determine oil quality parameters. It is used as a unique quantitative method to determine the acid value in CO, SFO, RPO, PNO, SBO, and SSO. UV–Vis spectrometer (320–800 nm) was used to acquire spectra of oil samples, along with principle component regression (PCR) and partial least squares regression (PLSR), for the quantitative analysis of spectral data. Calibration models were developed and subsequently validated. The correlation coefficient was found to range between 0.9197 and 0.9969 [26]. Additionally, UV–Vis spectroscopy is reported to have been used to determine fatty acid, diacylglycerol, and the amount of pigments in olive oils. The spectrum was obtained in the spectral range of 200–800 nm and was then used to construct prediction models for the chemical parameters. In total, 89 virgin olive oil (VOO) samples were studied and on application of PLS chemometrics, a correlation coefficient ranging between 0.5 and 0.8 was observed [27]. In another study, UV–Vis spectroscopy (200–800 nm) was used to estimate the acid value in sesame oil. The predicted and measured values were compared and a Pearson's linear correlation of 0.96 was recorded. In addition, the experimental results indicated that it is possible to estimate the acid values in sesame oil using the absorbance values with an average error of 3% [28].

In addition to UV–Vis spectroscopy, IR spectroscopy has also been widely used to assess oil quality. The fatty acids present in VOO have been quantified by the help of NIR (12 000–4000 cm^{-1}) followed by application of PLSR. The results indicated excellent prediction of fatty acids using the aforementioned approach with R^2 of 0.99 [29]. A similar study has been reported on virgin and extra VOO (EVOO) [30]. Additionally, peroxide value along with amount of fatty acids are reported to have been predicted with NIR spectroscopy in the range 11 500–4000 cm^{-1} and PLSR. An excellent prediction with R^2 0.92 and 0.99 was found for peroxide values and fatty acids, respectively [31]. In another study, Visible-NIR spectroscopy (400–2500 nm) was also employed along with the PLSR multivariate technique to predict and quantify free fatty acids and peroxide values in EVOO. The prediction of free fatty acids was found to be satisfactory (R^2 0.68), while peroxide value could be predicted correctly with R^2 0.97 [32]. Apart from NIR being employed to predict oil quality parameters, the use of MIR has also been reported. MIR spectroscopy (4000–650 cm^{-1}) along with PLSR has been utilized to determine the major fatty acids in 64 oil samples of Turkish VOO. Excellent prediction was observed with correlation coefficients of 0.97, 0.94, and 0.87, and 0.91, 0.94, and 0.97 for linoleic, oleic, and palmitic acid, and saturated, mono-unsaturated, and poly-unsaturated fatty acids, respectively [33].

FTIR spectroscopy in conjugation with PLSR has been widely studied to determine edible oil quality parameters, including peroxide value, acid value, fatty acid profiles, total phenols, relative density, refractive index, color pigments, and other minor components [25]. Dogruer *et al* recently investigated FT-IR spectroscopy to assess free fatty acids, total phenols and the fatty acid profile in various oils. Using a PLSR model in the 4000–650 cm^{-1} spectral region, great prediction with R^2 values of 0.95–0.97 and 0.92–0.96 for free fatty acids and total phenols, respectively,

Table 11.1. A summary overview of various spectroscopic techniques used for estimation of physiochemical properties of edible oils.

Objective	Oil	Spectral range/methods	Multivariate analysis	Prediction statistics/results	References
UV–Vis spectroscopy					
Determination of acid value	CO, SFO, RPO, PNO, SBO, SSO	320–800 nm	PLS, PCR	R^2>0.9900 and RMSE<0.0400.	[26]
Determination of fatty acid, diacylglycerol, and pigments	VOO	200–800 nm	PLSR	Correlation coefficients ranging between 0.5 and 0.8.	[27]
Prediction of acid value	SSO	200–800 nm	–	Average errors of the predictions were 3%.	[28]
IR spectroscopy					
Prediction of fatty acids	VOO	NIR 12 000–4000 cm^{-1}	PLSR	Excellent prediction of fatty acids with R^2 of 0.99.	[29]
Prediction of fatty acids	EVOO, VOO	NIR 1000–1800 nm	PLSR	Excellent prediction of fatty acids with R^2 of 0.95.	[30]
Prediction of fatty acid and peroxide value	EVOO	NIR, 11 500–4000 cm^{-1}	PLSR	Excellent prediction of fatty aid and peroxide value with R^2 of 0.92 and 0.99, respectively.	[31]
Prediction of free fatty acids and peroxide values	EVOO	Visible-NIR 400–2500 nm	PLSR	Excellent prediction of peroxide value with R^2 of 0.97. Satisfactory prediction of free fatty acid with R^2 of 0.68.	[32]
Determination of major fatty acid in Turkish VOO	VOO	MIR 4000–650 cm^{-1}	PLSR	Excellent prediction of major fatty acids named palmitic (R^2: 0.87), oleic (R^2: 0.94), linoleic (R^2: 0.97), saturated (R^2: 0.91), mono-unsaturated (R^2: 0.94) and poly-unsaturated (R^2: 0.97).	[33]

FTIR					
Prediction of diacylglycerol, wax and FAME	OO	4000–650 cm^{-1}	PLSR	Prediction ranging between R^2: 0.71–0.93.	[27]
Prediction of amount of color pigments	OO	4000–650 cm^{-1}	PLSR	Prediction ranging between R^2: 0.87–0.99.	[33]
Determination of free fatty acids, total phenols and fatty acid profile in cold pressed oils	PSO, GSO, SSO, BCO	4000–650 cm^{-1}	PLSR	Prediction ranging between R^2: 0.56–0.99.	[34]
Determination of fatty acids and triacylglycerols	VOO	4000–600 cm^{-1}	PLSR	Prediction with high correlation of 0.98.	[35]
Determination of fatty acid profiles of different oils	CO, PNO, SFO, SFFO, CAO, CSO	4000–700 cm^{-1}	PLSR	Prediction ranging between R^2: 0.923–0.960.	[36]
Determination of fatty acid profiles	OO	4000–650 cm^{-1}	PLSR	Prediction ranging between R^2: 0.69–0.97.	[37]
Estimation of peroxide value	OO, SFO, RPO	4000–650 cm^{-1}	PLSR	Prediction with high correlation of 0.70.	[38]
Estimation of peroxide value	WNO	4000–600 cm^{-1}	PLSR	Prediction with high correlation of 0.98.	[39]
Measurement of refractive index and relative density of oils	CAO, SFO, SBO	4000–450 cm^{-1}	PLSR; SVM	Excellent prediction of refractive index with R^2 0.98 and 0.93 obtained by using PLSR and SVM, respectively. Excellent prediction of relative density with R^2 0.98 and 0.96 obtained by using PLSR and SVM, respectively.	[40]

(Continued)

Table 11.1. *(Continued)*

Objective	Oil	Spectral range/methods	Multivariate analysis	Prediction statistics/results	References
NMR					
Determination of the fatty acid profile	VO	^{1}HNMR	–	Fatty acid profile of vegetable oil was successfully estimated.	[41]
Prediction of total, free and esterified sterols in extra VOO	EVOO	^{1}HNMR, ^{31}PNMR	BASA	With 96% forecast precision, total, free, and esterified sterols are well predicted.	[42]
Prediction of volatile organic compounds	EVOO	^{1}HNMR	OPLS-DA	With 98% forecast precision, total, free, and esterified sterols are well predicted.	[43]
Estimation of iodine value	CO, SFO, PSO, SFFO, RPO, SBO, WNO, LNO	^{1}HNMR	–	–	[44]
Determination of amount of fatty acids	SFO	^{1}HNMR	SPSS	Prediction ranging between R^2: 0.923–0.993.	[45]
Fluorescence					
Prediction of fatty acids, peroxide value and chlorophyll level	VOO	FFFS	PCA, FDA, CCSWA and PLSR	Fatty acid, peroxide value and chlorophyll level of VOO samples are predicted with R^2 of 0.98, 0.96 and 0.89, respectively.	[46]
Determination of fatty acid and peroxide value	VOO	SFS	PLSR and N-PLS	Excellent prediction of fatty acid, Peroxide value R^2 of 0.93 and 0.96 respectively.	[47]

Determination of peroxide index	EVOO	FS-detection wavelengths ranging from 197 nm to 998, 763 nm.	PLSR	R^2 close to 0.9 and 0.7 for total tocopherol content and peroxide index.	[48]
Determination of acid value	SSO	Fluorescence spectra from 400 to 800 with excitation nm at 405 nm were recorded using a portable fluorometer.	–	Average errors of the predictions were 8%.	[28]

BCO-black cumin oil; BO-blend oil; CAO-Canola oil; CMO-Camellia oil; CNO-Coconut oil; CO-corn oil; CSO-Cottonseed oil; EVOO-Extra VOO; FSO-Flaxseed oil; GO-Garlic oil; GSO-Grapeseed oil; HNO-hazelnut oil; LNO-Linseed oil; MO-Maize oil; MSO-Mustard oil; PNO-Peanut oil; PO-Palm oil; POO-Pomace olive oil; PSO-Pumpkin seed oil; RPO-Rapeseed oil; RO-Rice oil; ROO-refined olive oil ;ROPO-refined olive pomace oil; SAO-Sweet almond oil; SBO-Soybean oil; SFO-sunflower oil; SFFO-Safflower oil; SO-Soya oil; SSO-Sesame oil; VCNO-Virgin coconut oil; VO-Vegetable oil; WGO-wheat germ oil; WNO-Walnut oil.

was observed. In addition, the R^2 values for fatty acid profile using PLSR calibration were good. However, the prediction using separate validation sets was poor for various fatty acids, including heptadecanoic (R^2: 0.56) and behenic (R^2: 0.72), indicating that the established calibration model was overfitting. With the exception of these two models, all other PLSR models demonstrated dependability and robustness as measured by R^2 and RMSE [34].

FTIR spectroscopy (4000–650 cm^{-1}) along with PLSR has been used to determine wax, diacylglycerol, and fatty acid ethyl ester (FAEE) in olive oil. Besides being a quality parameter, FAEE is also considered to be an indicator of adulteration because it indicates the purity of oil. The predicted values for diacylglycerols, wax, and FAEE were found to have a high correlation of 0.71, 0.71, and 0.88–0.96, respectively [27]. The same group of researchers predicted the amount of color pigments in olive oil using FTIR in conjugation with PLSR, and obtained a high correlation ranging between 0.87 and 0.99 [33]. Fatty acids and triacylglycerols in VOO are also reported to have been determined using FTIR spectroscopy and PLSR. The models obtained a high correlation value of 0.98 [35]. In other studies, the fatty acid profiles of edible oils were found using FTIR spectroscopy along with PLSR chemometrics [36, 37]. Additionally, the peroxide values in olive, sunflower, rapeseed, and virgin walnut oil were determined using FTIR 4000–650 cm^{-1} and PLSR [38, 39]. In another study, the refractive index and relative density of canola, sunflower, and soybean oil are also reported to have been measured by FTIR spectroscopy coupled with PLSR and support vector machine (SVM) regression. The prediction results obtained by PLSR was better as compared to SVM for both refractive index (R^2: 0.98/0.93) and relative density (R^2: 0.98/0.96) [40].

In addition to the use of UV–Vis and IR spectroscopy, NMR spectroscopy also finds utility in determination of various oil quality parameters. ^{1}HNMR spectroscopy has been reported to be used for estimation of the fatty acid profile of vegetable oil [41]. In another study, ^{1}HNMR and ^{31}PNMR coupled with Bland and Altman statistical analysis (BASA) has successfully predicted the total, free, and esterified sterols in EVOO with a high predictive accuracy of 96% [42]. In addition, 1HNMR along with partial least squares-discriminant analysis (PLS-DA) has been employed to predict and quantify the volatile organic compounds in EVOO with an accuracy of 98% [43]. Moreover, the iodine value and the amount of fatty acids in various edible oil was also determined using NMR spectroscopy [44, 45].

Fluorescence spectroscopy has also been reported in the determination of various physiochemical properties of edible oil. The amount of fatty acids, peroxide values, and chlorophyll content of VOO was determined using front-face fluorescence spectroscopy (FFFS) along with PLSR chemometrics. The mathematical models were able to predict the high correlation of 0.98, 0.96, and 0.89 for fatty acid, peroxide value, and chlorophyll levels, respectively [46]. A synonymous study was conducted on VOO by applying PLSR and N-PLS to fluorescence spectra [47, 48]. In another study, the acid value of sesame oil was determined using fluorescence spectroscopy. A portable fluorometer was used to capture the fluorescence bands from 400 to 800 nm, with excitation at 405 nm. The experimental results indicated that forecast of acid values was possible by determination of absorbance values with an average error of 8% [28].

11.3.2 Detection of bioactive components

Oils have essential bioactive components, such as linoleic, oleic acids, tocopherol, squalene, biophenols, and phenolic compounds, which play critical roles in human health [19]. Tocopherol is an essential fat-soluble vitamin with strong antioxidant properties. It helps to preserve intracellular membrane integrity by ensuring physical and biochemical stability and by offering defense against oxidative tissue damage. The amount of tocopherol in edible oil can be predicted by spectroscopic techniques and multivariate statistical analysis (table 11.2). Gonçalves *et al* reported that the UV spectra obtained at 300–540 nm for corn, sunflower, soybean and olive oil can be successfully used to predict and evaluate tocopherols [49]. In another study, NIR (12 000–4000 cm^{-1}) spectroscopy was found to predict the amount of tocopherols in VOO with a high prediction correlation coefficient of 0.83 [29]. A synonymous study was conducted on vegetable oil and EVOO oil using surface enhanced Raman spectroscopy and fluorescence spectroscopy along with PLSR [48, 50].

Squalene is a highly unsaturated, unconjugated triterpene found largely in shark liver oils and, to a smaller extent, in plants such as olives. It has a number of therapeutic uses, e.g., it promotes blood circulation and has antioxidant, anti-inflammatory, and cell-rejuvenation properties [51]. With respect to quality of olive oil, squalene has been reported to reduce the peroxide value and provides an antioxidant action, and hence its presence needs to be estimated. It has been reported that squalene, and different components of the sterol and terpene portions may be predicted using Raman spectroscopy combined with Fourier transform with Raman scattered bands obtained at 1090–3090 cm^{-1} to predict squalene in olive oil [52].

The phenolic components in VOO have also been estimated. Numerous phenolic compounds are present in VOO, mostly hydroxytyrosol and its secoiridoid derivatives, oleuropein, and oleuropein aglycones. These polyphenolic components impart favorable biological effects, including the prevention of cancer and cardiovascular diseases. Thus, quantification of the polyphenolic components in oils is desirable and, as mentioned earlier, is feasible by employing different spectroscopic techniques. Uncu and Ozen reported the use of MIR spectroscopy (4000–650 cm^{-1}) in conjugation with PLSR to detect hydroxytyrosol as major phenolic compound and the total phenolic content in VOO. This technique was found to be highly reliable and a high correlation of 0.97 and 0.99 was observed for hydroxytyrosol and total phenolic content, respectively [33]. In a similar study, Squeo *et al* observed excellent prediction (R^2 0.95) of phenolic level in VOO by applying PLSR to fluorescence spectra [53].

11.3.3 Geographical origin traceability

Although the oil may be of comparable quality, oils produced from oilseeds procured from certain specific nations are regarded as more valuable than those from other countries [54]. As a result, product origin characterization and

Table 11.2. A summary overview of various spectroscopic techniques that are used to estimate the bioactive components in edible oils.

Objective	Oil	Method	Multivariate analysis	Prediction statistics/results	References
Evaluate tocopherols in VOO	VOO	NIR 12 000–4000 cm-1	PLSR	Good prediction of tocopherols with R^2 of 0.83.	[29]
Evaluate tocopherols in extra VOO	EVOO	FS-detection wavelengths ranging from 197 nm to 998, 763 nm.	PLSR	Prediction with high correlation of nearly 0.9.	[48]
Evaluate tocopherols in oil	CO, SFO, SBO, OO	UV-300 to 540 nm	MCR-ALS	—	[49]
Estimation of α-tocopherol in vegetable oil	VO	MIPs-SERS (600–1800 cm^{-1})	PCA and PLS	R^2 greater than 0.92 and RSME less than 0.41 for total tocopherol content prediction.	[50]
Determination of squalene content	OO	FT-RS (1090–3090 cm^{-1})	MDS	FT-RS has been used successfully to determine the spectrum of unsaponifiable matter in varietal olive oils as well as other refined and crude edible oils.	[52]
Determination of phenolic compounds in oil	VOO	MIR 4000_650 cm^{-1}	PLSR	A high correlation of 0.97 and 0.99 was observed for hydroxytyrosol and total phenolic content was found	[33]
Prediction of phenolic level	VOO	FFFS	PLSR	Excellent prediction of phenolic level since R^2 of 0.95 was observed.	[53]

BCO-black cumin oil; BO-blend oil; CAO-Canola oil; CMO-Camellia oil; CNO-Coconut oil; CO-corn oil; CSO-Cottonseed oil; EVOO-Extra VOO; FSO-Flaxseed oil; GO-Garlic oil; GSO-Grapeseed oil; HNO-hazelnut oil; LNO-Linseed oil; MO-Maize oil; MSO-Mustard oil; PNO-Peanut oil; PO-Palm oil; POO-Pomace olive oil; PSO-Pumpkin seed oil; RPO-Rapeseed oil; RO-Rice oil; ROO-refined olive oil ;ROPO-refined olive pomace oil; SAO-Sweet almond oil; SBO-Soybean oil; SFO-sunflower oil; SFFO-Safflower oil; SO-Soya oil; SSO-Sesame oil; VCNO-Virgin coconut oil; VO-Vegetable oil; WGO-wheat germ oil; WNO-Walnut oil.

certification are critical, especially given the growing number of oil producing nations and the global interest in edible oils. Currently, this special concern is also focused on evaluating the authenticity of reported geographical origin, in addition to its safety and quality management [55]. It should also be noted that food traceability has an influence on customer trust in product quality control, specific suppliers and geographical origin evaluation. Inevitably, an instance of fraud can have substantial effects, affecting both individual customers and the entire market [56]. As a result, the ongoing regulatory revision pertaining to oil quality intends to facilitate product marketing, promote true labeling, and set standard rules for proper claims [57]. The grounds for evaluating the geographical origin of various edible oils are mostly connected to necessary laws and/or to the necessity of truthful branding and labeling. Moreover, as previously stated, falsifying the varietal and/or geographical origin of different oils is a kind of fraud. Therefore, international organizations have devised a significant piece of law, the Protected Designation of Origin (PDO), to maintain its quality [58]. This law also enables the recognition of specific EVOOs as per their geographical areas from where they originate.

The geographical origin of EVOO can be traced using different spectroscopic techniques. However, regardless of the technique used, the usual approach for its determination for PDO olive oils is to analyze numerous oil samples from varied geographical areas and then use the chemometrics tool to construct a prediction or classification model for unknown samples (table 11.3) [59]. In 2012, Casale *et al* conducted a study for geographical validation of oil samples collected from three different Italian regions using UV–Vis spectroscopy (190–1100 nm), principal component analysis (PCA), and soft independent modeling by class analogy (SIMCA). Excellent differentiation of EVOO based on geographical origin with 100% accurate categorization was obtained [60]. A similar study was conducted for validation of EVOO samples procured from three Spanish regions. This study employed UV–Vis spectroscopy along with LDA and PLS-DA to discriminate oil samples. This technique could correctly classification 92.5% of the samples [61].

IR spectroscopy has also been utilized for geographical validation of EVOO samples. Lin *et al* employed NIR (800–1075 nm) spectroscopy along with PLS chemometrics for geographical authentication of oil samples originating from Spanish, Turkish, and Italian regions. This method could successfully discriminate the oil samples and R^2 of 0.987 was observed [62]. In another study, geographical verification of Ligurian and non-Ligurian oil samples from Italy was accomplished using NIR spectroscopy (1100–2498 nm). This method could classify the oil samples correctly, with 81.5% and 82.2% correct classification as observed on subjecting the obtained spectra to PLS-DA and SVM, respectively [63]. Similar studies have been conducted on oil samples from Moroccan, Greece, Portuguese, Spanish and Italian regions [46, 60, 64]. Even Raman spectroscopy is reported to have been used for authentication of EVOO from Italian and Spanish region [65, 66].

NMR spectroscopy has also been used for geographical origin traceability of EVOO samples. Longobardi *et al* used ^{1}HNMR for geographical origin authentication of oil samples from various regions in Greece and Italy. The spectra were analyzed by PCA coupled with CA and NCM approach, and it was observed that

Table 11.3. A summary overview of various spectroscopic techniques used for geographical origin traceability of edible oils

Objective	Oil	Method	Multivariate analysis	Prediction statistics/results	References
UV–Vis spectroscopy					
Geographical authentication of oil samples from three Italian regions	EVOO	190–1100 nm	PCA, UNEQ and SIMCA	Excellent discrimination of oil samples, 100% of correct classification was obtained.	[60]
Geographical authentication of oil samples from three Spanish regions	EVOO	190–1090 nm	LDA and PLS-DA	Excellent discrimination of oil samples, 92.5% of correct classification was obtained.	[61]
IR spectroscopy					
Geographical authentication of oil samples from Turkish, Spanish, and Italian regions	EVOO	NIR 800 and 1075 nm	DOSC-GA-PLS	Excellent discrimination of oil samples, R^2 of 0.987 was observed.	[62]
Geographical authentication of Ligurian and non-Ligurian oil samples	EVOO	NIR 1100–2498 nm	PLS-DA and SVM	Good classification of oil samples, with 85.1% and 82.2% of correct classification obtained by applying SVM and PLS-DA, respectively.	[63]
Geographical authentication of oil samples from three Italian regions	EVOO	NIR 10 000–4000 cm^{-1}	PCA, UNEQ and SIMCA	Excellent discrimination of oil samples was observed.	[60]
Geographical authentication of oil samples from five Moroccan regions	VOO	MIR 4000–700 cm^{-1}	PCA, FDA and PLSR	Excellent discrimination of oil samples, 91.87% of correct classification was obtained.	[46]
Geographical authentication of oil samples from Greece, Italy, Portugual, and Spain	EVOO	FTIR 800–4000 cm^{-1}	PLS, LDA, GA	Good classification of oil samples with 96% and 100% of correct classification obtained by applying PLS-LDA and GA-LDA, respectively.	[64]
Raman spectroscopy					
Geographical authentication of oil samples from seven Italian regions	EVOO	1800–900 cm^{-1}	PCA and LDA	Samples are clearly discriminated with 82 % of correct classification.	[64]

Geographical authentication of oil samples from Spanish regions	EVOO	3100–100 cm^{-1}	PCA and FDA	Samples are clearly discriminated with 89 % of correct classification.	[66]
NMR Spectroscopy					
Geographical authentication of oil samples from Greece and Italy	EVOO	^{1}HNMR	PCA and PCACA-NCM	Correct prediction probabilities of 78% were achieved with region-specific correct predictions ranging between 53% and 100%.	[67]
Geographical authentication of oil samples from Malta	EVOO	^{1}HNMR, ^{13}CNMR	PCA, PLS-DA, ANN	Satisfactory level of discrimination was observed.	[68]
Geographical authentication of oil samples from Italy, Greece, and Spain	EVOO	^{1}HNMR	PCA, CA, KNN	Excellent discrimination of oil samples as 96% of correct classification was obtained.	[69]
Geographical authentication of international Blends with oil from Italy, Tunisia, Portugal, Spain, and Greece	EVOO	^{1}HNMR	PCA, PLSR, OPLS-DA	–	[70]
Fluorescence spectroscopy					
Geographical authentication of oil samples from seven Argentinian regions	EVOO	EEFS	PCA and PLS-DA	Excellent discrimination of oil samples, 100% of correct classification was obtained.	[71]
Geographical authentication of oil samples from two Maltese regions	EVOO	EEFS	PCA, PARAFAC, LDA, DN-PLSR	Clear discrimination of oil samples as 93%, 80%, and 73% of correct classification was obtained by applying, DN-PLS, PARAFAC, and LDA, respectively.	[72]
Geographical authentication of oil samples from five Italian regions	EVOO	FFFS	PCA, LDA, SVM and k-NN	Excellent discrimination of oil samples, 94% of correct classification was obtained.	[73]
Geographical authentication of oil samples from five Moroccan regions	VOO	FFFS	PCA, FDA, CCSWA and PLSR	Excellent discrimination of oil samples,96.72% of correct classification was obtained.	[46]

BCO-black cumin oil; BO-blend oil; CAO-Canola oil; CMO-Camellia oil; CNO-Coconut oil; CO-corn oil; CSO-Cottonseed oil; EVOO-Extra VOO; FSO-Flaxseed oil; GO-Garlic oil; GSO-Grapeseed oil; HNO-hazelnut oil; LNO-Linseed oil; MO-Maize oil; MSO-Mustard oil; PNO-Peanut oil; PO-Palm oil; POO-Pomace olive oil; PSO-Pumpkin seed oil; RPO-Rapeseed oil; RO-Rice oil; ROO-refined olive oil ;ROPO-refined olive pomace oil; SAO-Sweet almond oil; SBO-Soybean oil; SFO-sunflower oil; SFFO-Safflower oil; SO-Soya oil; SSO-Sesame oil; VCNO-Virgin coconut oil; VO-Vegetable oil; WGO-wheat germ oil; WNO-Walnut oil.

the former approach allowed better differentiation between the oil samples as compared to PCA alone. Furthermore, accurate forecasts of 78% were found, with region-specific predictions ranging from 53% to 100% [67]. Another study depicted the utilization of 1H and 13C NMR technique for authentication of 65 oil samples from Maltese (30) and foreign places (35). The spectra were further analyzed by PCA, PLS-DA, and artificial neural network (ANN), and the models showed satisfactory discrimination of oils [68]. In addition, Winkelmann and Küchler used ^{1}HNMR for the distinction of olive oil samples from various regions in Italy, Greece, and Spain. The geographical origin was authenticated using PCA, CA, and the K-Nearest Neighbors (KNN) algorithm, and 96% of correct classification was observed [69]. Moreover, Calo *et al* used conventional and multi-suppressed ^{1}H-NMR tests to investigate marketed international EVOO mixes from Greece, Italy, Portugal, Spain, and Tunisia. The probable relationship between the features of the EVOOs blends (as described by their ^{1}H-NMR profiles) and the blend composition was explored. For the analyzed dataset, the characteristics of international blends were substantially linked with the proportion of Italian EVOO ingredients, suggesting a probable effect of Coratina-based oils [70].

The authentication of EVOO based on its geographical origin has also been accomplished by fluorescence spectroscopy. Jimenez-Carvelo *et al* employed excitation-emission fluorescence spectroscopy for the differentiation of 65 oil samples as per their eight geographical origins. PCA and PLSDA was applied to the spectra and 100% of correct classification of samples as per their geographical origin were observed [71]. In addition, geographic origin traceability of oil samples from Maltese regions were also classified by using excitation-emission fluorescence spectroscopy coupled with LDA, PARAllel FACtor (PARAFAC) analysis, and DN-PLS with 73%, 80%, and 93% correct classification, respectively [72]. Similar studies of the geographical validation of oil samples from Italian and Moroccan regions using FFFS have also been published [46, 73].

11.3.4 Discrimination among different cultivars

The olive cultivar used for processing olive oil is important for appropriately assigning the mark of authenticity because botanical origin is one of the conditions for PDO assignment [74]. Therefore, discrimination among different cultivars is essential, and spectroscopic techniques have been widely used for this purpose (table 11.4). Laroussi-Mezghani *et al* employed NIR spectroscopy in the spectral range of 10 000–4500 cm^{-1} for verification of six Tunisian varieties EVOO. This study used PCA, SNV, and SIMCA for its statistical analysis, and this method could successfully separate samples with 98.5% of correct classification [75]. The outcome was in agreement to another study conducted on authentication of olive oil of three Italian varieties using MIR spectroscopy (4000–400 cm^{-1}) and LDA-SIMCA chemometrics. Excellent sample separation with 98% and 100% precision categorization was observed [76]. A synonymous study was performed by Zaroual *et al*, who successfully distinguished 41 Moroccan VOO based on their geographical origin and variety, with 91.78 and 91.78% right classification, respectively [46]. In

Table 11.4. A summary overview of various spectroscopic techniques used for cultivar discrimination.

Objective	Oil	Method	Multivariate analysis	Prediction statistics/results	References
Varietal authentication of six Tunisian varieties	EVOO	NIR 10 000–4500 cm^{-1}	PCA, SNV and SIMCA	Excellent varietal discrimination of oil samples, a correct classification of 98.5% was obtained.	[75]
Varietal authentication of three Italian varieties	EVOO	MIR 4000–400 cm^{-1}	LDA and SIMCA	Excellent varietal discrimination of oil samples, a correct classification of 98%–100% was obtained.	[76]
Varietal authentication of four cultivars	VOO	MIR 4000–700 cm^{-1}	PCA, FDA and PLSR	Good varietal discrimination of oil samples, a correct classification was 91.8% was obtained.	[46]
Botanical discrimination of Greek olive oils	EVOO	ATR-FTIR spectroscopy	Stepwise LDA analysis and QDA models	Correct classification score of 97.4% and 100.0% was obtained using LDA and QDA, respectively.	[77]
Varietal authentication of four cultivars (Frantoio, Hojiblanca, Picual, and Gorda)	EVOO	Raman 3100–100 cm^{-1}	PCA and FDA	Samples are clearly discriminated with 84.0% of correct cultivar classification.	[66]
Varietal authentication of cultivars (Gemlik, Ayvalık, Memecik, and Bienchera)	EVOO	^{1}HNMR	PCA and PLS-DA	Excellent varietal discrimination of oil samples, a correct classification of 92.5% was obtained.	[29]
Varietal authentication of five cultivars (Coratina, Dritta, Grossa di CassanoMoraiolo, and Picholine)	EVOO	^{13}CNMR	PCA, PLS, PCR, MLR	70%–100% correct varietal classification was observed.	[78]
Varietal authentication of four cultivars	VOO	FFFS	PCA, FDA, CCSWA and PLSR	Good varietal discrimination of oil samples, a correct classification of 95.12% was obtained.	[46]
Varietal authentication of four Italian varieties	EVOO	FFFS	PCA, LDA, SVM and k-NN	Samples are clearly discriminated with 90.0%, of correct cultivar classification.	[73]

BCO-black cumin oil; BO-blend oil; CAO-Canola oil; CMO-Camellia oil; CNO-Coconut oil; CO-corn oil; CSO-Cottonseed oil; EVOO-Extra VOO; FSO-Flaxseed oil; GO-Garlic oil; GSO-Grapeseed oil; HNO-hazelnut oil; LNO-Linseed oil; MO-Maize oil; MSO-Mustard oil; PNO-Peanut oil; PO-Palm oil; POO-Pomace olive oil; PSO-Pumpkin seed oil; RPO-Rapeseed oil; RO-Rice oil; ROO-refined olive oil ;ROPO-refined olive pomace oil; SAO-Sweet almond oil; SBO-Soybean oil; SFO-sunflower oil; SFFO-Safflower oil; SO-Soya oil; SSO-Sesame oil; VCNO-Virgin coconut oil; VO-Vegetable oil; WGO-wheat germ oil; WNO-Walnut oil.

addition, ATR-FTIR reported to have been used for botanical differentiation of Greek EVOO. Accurate classification score derived from linear and quadratic discriminant cross-validation analysis was 97.4% and 100.0%, respectively [77].

Raman, NMR, and fluorescence spectroscopy have also been used for discrimination of olive oil on the basis of its cultivar. Sanchez-Lopez *et al* showed that Raman spectroscopy (3100–100 cm^{-1}) coupled with PCA and FDA could clearly discriminate between Frantoio, Gorda, Hojiblanca, and Picual varieties of olive oil with correct classification score of 84% [66]. ^{1}HNMR spectroscopy was employed for varietal classification and validation of EVOO of Ayvalık, Bienchera, Gemlik, and Memecik varieties. Remarkable distinction of the varieties was achieved and 92.5% of precise characterization was observed [29]. In another study, ^{13}CNMR was used for classification of olive oil samples from three cultivars produced in the region of Tuscany. Using PCA, PLS, and PCR, 100% correct varietal prediction was observed. Synonymous research was carried out by Shaw *et al* [78]. Moreover, varietal authentication of VOO is also feasible by using fluorescence spectroscopy. FFFS in conjugation with PCA, FDA, and PLSR was used and good discrimination of samples with 95.12% correct classification was observed [46]. An analogous study was undertaken for cultivar verification of EVOO from Italian regions [73].

11.3.5 Detection of adulteration

Food adulteration is defined as the deliberate depreciation of the food quality, either by mixing or replacement of inferior components, or through the elimination of a valuable element [79]. Edible oil is important nutritionally and economically, and therefore its high demand encourages adulteration. Adulteration not only degrades the quality of edible oil but can also sometimes be a health hazard. Edible oils and fats are adulterated in different ways, of which adulteration of cold pressed oil with refined oil and admixing of expensive oil with less expensive oil are very common. Cold pressed vegetable oils are in high demand since they are generated solely by pressing and basic filtering without processing. As a result, unlike refined oils, cold press oil has no chemical–solvent interaction. In addition, undesirable compounds such as trans fatty acid, which increases the risk of heart disease, are absent in cold pressed oils [80]. Trans fatty acids are produced during the refining process by partial hydrogenation and are more detrimental than those that are naturally present [81]. Nonetheless, given that cold pressed oil is more costly as compared to refined oil, it is common for them to be adulterated with the economical refined oils.

Some oils are at greater risk of being adulterated because of their high prices and confined accessibility—expensive oil is substituted with a cheap oil of similar fatty acid profile to increase the profit margins. For example, comparatively less expensive vegetable oils (canola, rapeseed, and mustard seed oil) are admixed with olive oil [82]. Moreover, sesame oil, due to its high cost, is adulterated with sunflower, corn, and other cheap vegetable oil [83]. Vegetable oil (rice bran oil), which is cheaper than mustard oil, is mixed with artificial allyl isothiocyanate along with oil soluble yellow dye to pass for mustard oil [84]. In addition, Navya *et al* reported karanja oil, castor oil, mineral oil, and artificial colors are the major oil adulterants [85]. Argemone oil is also recognized as the major adulterant of mustard oil. These adulterants have detrimental health effects. Argemone oil in edible oil can cause glaucoma, epidemic dropsy, and loss of eyesight [86]. Moreover, vegetable oil adulterated with mineral oil may lead to liver damage and have carcinogenic effects, while castor oil adulterated oils can cause stomach problems [85]. Hence, the edible oils available in markets need to be monitored and checked for adulteration.

Various standard qualitative chemical tests are employed to detect admixing in edible oils. The majority of these approaches are based on the appearance of turbidity, creation of precipitate or development of characteristic color. However, analytical techniques can be applied to detect the level of impurity in various oil samples. The components of oils are generally studied using different analytical techniques to detect their adulteration. Because each oil has a specific component at a particular level, it serves as a feasible detection parameter [87]. In addition, the fatty acid profile of oils can be used to detect adulteration because each oil has a different fatty acid profile. Spectroscopic techniques serve as effective tools to detect adulteration in edible oils (table 11.5).

Aroca-Santos *et al* reported the use of UV–Vis spectroscopy to detect the admixing of soybean and sweet almond oils in pure EVOO. The spectra in 380–800 nm spectral range for 141 oil samples were then subjected to PLSR and MLP chemometrics. This technique could satisfactorily detect the presence of adulterants with 2% as minimum detection limit and mean absolute error of 2.84% [88]. In another study, the adulteration of EVOO with refined/pomace olive, corn, and sunflower oil was successfully ascertained using UV–Vis spectroscopy (190–900 nm) and PLSR. The results also indicated that adulteration was detected at a level as low as 5% with correlation coefficient and RMSEP of 0.97 and 0.007% [89]. A similar study has been reported on detection of sunflower, corn, and soybean oil admixing in EVOO [90].

Several studies have also reported the use of IR and FTIR spectroscopy for detection of admixing in various edible oils. Yuan *et al* reported the application of FTIR-NIR (780–2526 nm) and PLS for detection of admixing in camellia oil with soybean oil and maize oil in the range of 0%–50% and 0%–40%, respectively [91]. In another study, camellia oil admixed with 1%–100% of sunflower oil, corn oil, or peanut oil were analyzed using NIR spectroscopy in the spectra range of 1000–2500 nm. By applying PLS to the spectra, a high correlation (0.9988) and low root mean square error (1.52%) was observed for this technique [92]. Moreover, numerous studies have been reported on adulteration detection in olive oil using IR and elated spectroscopic techniques. Christy *et al* proposed application of NIR spectroscopy along with chemometrics (PLS) to detect blending of olive oil with coconut, soya, sunflower, hazelnut, and walnut oil [93]. Oils contain distinct fatty acid composition, and hence depict different bands in the resultant spectra, and this forms the basis to detect presence of foreign oils. The results indicated a high correlation value for each adulterant. Another study separated pure EVOO samples from oil samples mixed with mild deodorized or refined olive oil samples, using NIR (10 000–4400 cm^{-1}) spectroscopy and PLSR technique [38]. Similar studies were conducted on EVOO for detection of admixing with different vegetable oils [94–106]. Moreover, adulteration of corn, rapeseed, rice, peanut, and blend oil in sesame

Table 11.5. A summary overview of various spectroscopic techniques used for detection of adulterants in edible oils.

Pure oil	Adulterants	Spectral range used/methods	Detection range (%)	Multivariate analysis	Prediction statistics	References
UV–Vis spectroscopy						
EVOO	SBO, SAO	380–800 nm	2%–20%	PLSR and MLP	Detection of the adulteration at a level of 1% with mean absolute error of 2.84%.	[88]
EVOO	ROO, ROPO, SFO, CO	190–900 nm	< 10%	PLSR	Detection of the adulteration at a level of 5% with R^2 of 0.97 and RMSEP of 0.007%.	[89]
EVOO	CO, SBO, SFO	200–400 nm	0%–5%	PCA, PLSR	Detection of the adulteration with R^2 of 0.863, 1.000 and 0.981 and RMSEP of 0.71%., 0.001% and 0.271% for EVOO-corn oil, EVOO-soybean oil and EVOO-sunflower oil blends, respectively.	[90]
IR spectroscopy						
CMO	SBO, MO	FTIR-NIR 780 to 2526 nm	0%–50%; 0%–40%	PLS	Detection of the adulteration with correlation coefficient of 0.999 96 and 0.999 61 for soybean and maize oil blends.	[91]
Camellia oil	SFO, CO, PNO	NIR 1000–2500nm	1%–100%	PLS	Detection of the adulteration with correlation coefficient and mean absolute error of 0.9988 and 1.52%.	[92]
Olive oil	CNO, SO, SFO, HNO, WNO	FT-NIR 12 000–4000 cm^{-1}	0–100	PCA; PLS	Detection of the adulteration with correlation coefficient of 0.9997, 0.9978, 0.9985, 0.9994,and 0.9994 for CNO, SO, SFO, HNO, and WNO, respectively.	[93]
EVOO	Deodorized and ROO	NIR 10 000–4400 cm^{-1}	2.5%–75%	PLSR and PCR	Detection of the adulteration ranging between 2.5% and 75% with R^2 of 0.97and 0.98 obtained with PLSR and PCR, respectively.	[38]

EVOO	CO, SFO, SBO, CAO	NIR 12 000–4000 cm^{-1}	10% and 20%	PCA	Detection of the adulteration ranging between 10% and 20% with clear differentiation of samples.	[94]
EVOO	CAO, SFO	NIR 12 000–4000 cm^{-1}	2.5%–50%	BSS-PLS	Detection of the adulteration with R^2 of 0.99 and root mean square error of 1.44%.	[95]
EVOO	CAO, SBO, CO	NIR 12 000–4000 cm^{-1}	5%–95%	2D-COS and k-NN	Detection of CAO, SBO, and CO with 85% accuracy.	[96]
EVOO	SFO	MIR 4000–600 cm^{-1}	1% and 24 %	PLSR	With less than 2% prediction error and an R^2 of 0.99, excellent detection of adulteration was observed.	[97]
EVOO	PO	MIR4000–500 cm^{-1}	1% and 50%	PLSR and PCR	With R^2 0.99, pure EVOO samples were clearly distinguished from admixed ones.	[98]
EVOO	GSO, WNO, SBO	FTIR 3018–3002, 1200–1000 cm^{-1}	1%–50%	PLSR	Detection of the adulteration with correlation coefficient and root mean square error of 0.99 and 1.32%.	[99]
EVOO	CMO, SBO SFO, CO	FTIR 4000–400 cm^{-1}	1%–90%	PLSR	Detection of the adulteration with correlation coefficient ranging between 0.97 and 0.99	[100]
EVOO	PNO	FTIR 3050–2750, 1800–600 cm^{-1}	0.5%–5%, 6%–30%	PLSR	Detection of the adulteration with correlation coefficient and root mean square error of 0.977 and 1.04%.	[101]
EVOO	SBO	FT-MIR 4000–350 cm^{-1}	40%–65%	PLSR	Detection of the adulteration with correlation coefficient and root mean square error of 0.98 and 4.89%.	[102]
EVOO	CO, SFO	FTIR 3027–3000, 1076–860, 790–698 cm^{-1}; 3012–3000 cm^{-1}	1%–50%	PLSR	Detection of the adulteration with correlation coefficient and root mean square error of 0.99 and 1.13% and 0.98 and 2.02% for corn and sunflower oil, respectively.	[103]

(Continued)

Table 11.5. (*Continued*)

Pure oil	Adulterants	Spectral range used/methods	Detection range (%)	Multivariate analysis	Prediction statistics	References
EVOO	RBO	FTIR 3020–3000, 1200–900 cm^{-1}	1%–50%	PLSR	Detection of the adulteration with correlation coefficient and root mean square error of 0.981 and 2.15%.	[104]
EVOO	SSO	FTIR 3050–2927, 1517–1222, 1207–1018 cm^{-1}	5%–60%	PLSR	Detection of the adulteration with correlation coefficient and root mean square error of 0.99 and 1.01%.	[105]
EVOO	HNO, CAO, SFO, PMO	FTIR 3805–2841, 1877–1105 cm^{-1}	5%–40%	PLSR	Detection of the adulteration with R^2>0.90.	[106]
SSO	CO, RO, PNO, RPO, BO	NIR10,000–4500 cm^{-1}	0–100	CARS; ECR; PLS	Detection of the adulteration with root mean square error of calibration and prediction of 0.0188 and 0.0398.	[107]
Virgin CNO	CO, SFO	FTIR 4000–650 cm^{-1}	2%–100%; 0%–100%	PLS; DA	Excellent detection of the adulteration was observed as the correlation coefficient was 0.999.	[108]
VCNO	MSO	FTIR 1800–720 cm^{-1}	1%–13%	PLSR	Detection of the adulteration with correlation coefficient and root mean square error of 0.999 and 0.125%.	[109]
	GSO, SBO	FTIR 3027–2985, 1200–900 cm^{-1}; 3025–2995, 1200–1000 cm^{-1}	1%–50%	PLSR	Detection of the adulteration with correlation coefficient and root mean square error of 0.994 and 1.32% and 0.996 and 1.04%, for GSO and SBO, respectively.	[110]
	OO, PO	FTIR 1120–1105, 965–960 cm^{-1}	1%–50%	PLSR	Detection of the adulteration with correlation coefficient and root mean square error of 0.997 and 0.823%.	[98]
MSO	AGO	FTIR 1800–500 cm^{-1}	1%–30%	PLSR	Detection of the adulteration with correlation coefficient and root mean square error of 0.999 and 0.2%.	[111]

Raman spectroscopy						
Olive oil	RPO, SSO, GO, SFO, WNO, SFFO, SBO, WGO, FSO	700–3100 cm^{-1}	5%–100%	PLS; PCA	Detection of the adulteration with correlation coefficient and root mean square error of 0.971 and 0.036%.	[112]
WNO, PSO	SFO	200–2000 cm^{-1}	2.5–50	PLS	Correct prediction of admixing with prediction degree >95 %.	[113]
EVOO	SBO, SFO, CO	200–2000 cm^{-1}	1%–10%, 1%–10%, 2%–98%	ESM and SVM	Detection of the adulteration with correlation coefficient and root mean square error of 0.9956, 0.9915 and 0.9905 for samples adulterated with SBO, CO, SFO, respectively. The mean absolute relative errors calculated using ESM and SVM are 7.41%, 7.78% and 9.45% and 5.10%, 6.96% and 4.55%, respectively.	[114]
EVOO	HNO	4000–550 cm^{-1}	1%–90%	CLPP, kNN	CLPP combined with kNN outperforms SIMCA and PLS-DA in classification performance.	[115]
EVOO	SBO	3500 to 50 cm^{-1}	40%–65%	PLSR	Detection of the adulteration with correlation coefficient and root mean square error of 0.99 and 1.57%.	[102]
NMR spectroscopy						
PNO	SBO, RPO, PO	^{1}HNMR	0–100	PCA	Correct classification of 91.7%, 79.2%and 87.5% for PEO + PAO, PEO + RO, and PEO + SO, respectively.	[116]
ROO	Refined HNO	^{1}HNMR and ^{31}PNMR	0%–100%	CBT, CDA	A high correlation of 0.96 and 0.80 for the result prediction was achieved using CBT and CDA, respectively.	[117]

(*Continued*)

Table 11.5. (*Continued*)

Pure oil	Adulterants	Spectral range used/methods	Detection range (%)	Multivariate analysis	Prediction statistics	References
EVOO	SFO, SBO, HNO, PNO	^{1}HNMR	10%–50%	LDA	Excellent identification of sunflower, SBO, HNO, and PNO in EVOO, with classification accuracy varying between 98% and 100%.	[118]
CMO	CO, SFO, RPO	^{1}HNMR	5%–80%	PCA, OPLS-DA, PLS	Excellent detection with discrimination accuracy greater than 90%.	[119]
Fluorescence spectroscopy						
EVOO	ROO	SFS	2% and 20%	PLSR	Detection with high accuracy as correlation of 0.94 was observed.	[120]
EVOO	POO	SFS	3% and 33%	LDA-PARAFAC, UPLS and DA-UPLS	Detection with high accuracy of 99%.	[121]
EVOO	SBO	FS	1% and 30%	PLSR	High accuracy of detection was observed as 0.98 correlation coefficient was recorded.	[122]
EVOO	SBO	LIF	3%–20%	PCA, LDA and PLSR	High accuracy of detection was observed as 0.92 and 0.27% correlation coefficient and root mean square error was recorded	[123]

BCO-black cumin oil; BO-blend oil; CAO-Canola oil; CMO-Camellia oil; CNO-Coconut oil; CO-corn oil; CSO-Cottonseed oil; EVOO-Extra VOO; FSO-Flaxseed oil; GO-Garlic oil; GSO-Grapeseed oil; HNO-hazelnut oil; LNO-Linseed oil; MO-Maize oil; MSO-Mustard oil; PNO-Peanut oil; PO-Palm oil; POO-Pomace olive oil; PSO-Pumpkin seed oil; RPO-Rapeseed oil; RO-Rice oil; ROO-refined olive oil ;ROPO-refined olive pomace oil; SAO-Sweet almond oil; SBO-Soybean oil; SFO-sunflower oil; SFFO-Safflower oil; SO-Soya oil; SSO-Sesame oil; VCNO-Virgin coconut oil; VO-Vegetable oil; WGO-wheat germ oil; WNO-Walnut oil.

oil (0%–100%) is reported to have been detected by NIR spectroscopy and competitive adaptive reweighted sampling (CARS), elastic component regression (ECR), and PLS techniques [107].

The adulteration of virgin coconut oil with corn oil and sunflower oil in 2%–100% and 0%–100% has also been detected using FTIR spectroscopy and PLS-DA. The results indicated high prediction levels with correlation of 0.999 [108]. Synonymous studies were reported on detection of mustard seed oil, rapeseed oil, grapeseed oil, palm oil, soybean oil, or olive oil admixed in pure virgin coconut oil samples [98, 109, 110]. In another study, Jamwal *et al* detected presence of 1to 30% of argemone oil in mustard oil samples using FTIR spectroscopy in the spectral range of 1800–500 cm^{-1}. The spectra obtained were then studied using PLSR and high correlation coefficient of 0.999 was observed [111].

Raman spectroscopy using visible spectrum illumination Raman spectroscopy with excitation in the visible spectral range (VIS Raman) is reported to be a potential method to detect admixing in olive oil [112]. Raman spectra (at 700–3100 cm^{-1}) were discovered and examined for various bean oils in this research. They found significant similarity between sunflower and olive oil, the main difference between them was observed in 800–1800 and 2850–3020 cm^{-1} range. It was also noticed that as the concentration of sunflower increases, the carotenoid bands at 1008, 1150 and 1525 cm^{-1} decreases. This study reported the detection of adulterants in 5%–100% with high correlation (0.971) and low root mean square error value (0.036) obtained by analyzing the spectra using PLS chemometrics. Becze and Simedru used Raman Spectroscopy along with PLS method to detect the adulteration in walnut oil and pumpkin oil [113]. In this study, pure walnut and pumpkin oils were adulterated with sunflower oil and Raman Spectra were found for all the samples. Then, the PLS model was used to design the model for the prediction of which the prediction level was found to be greater than 95%. Another research reported utilization of Raman spectroscopy (200–2000 cm^{-1}) to detect soybean oil, sunflower oil, and corn oil in EVOO in the admixing range of 1%–10%, 1%–10%, and 2%–98%. The spectra were further analyzed by ESM and SVM. The mean absolute relative errors calculated for soybean, corn and sunflower oil by ESM and SVM were 7.41%, 7.78% and 9.45% and 5.10%, 6.96% and 4.55%, respectively. It was observed that Raman vibrational band intensities were found to be faster and simpler than SVM methods [114]. Similar studies were conducted using Raman spectroscopy for detection of various vegetable oils in EVOO [102, 115].

The use of NMR spectroscopy for the detection of admixing in various oil samples has also been reported. Zhu *et al* employed low frequencies ^{1}HNMR and chemometrics (PCA) to detect admixing of soybean, rapeseed, and palm oil in peanut oil [116]. It was found that the level of admixed adulterants should be at least 10% to be successfully detected, and the results recorded 91.7%, 79.2%, and 87.5% of correct classification rate for palm oil, rapeseed oil, and soybean oil, respectively. In another study, Agiomyrgianaki *et al* employed 1H and 31P NMR to detect 0%–100% admixing of refined hazelnut oil to refined olive oil [117]. The results were analyzed by classification binary trees (CBT) and canonical discriminant analysis (CDA), and a high correlation of 0.96 and 0.80 for the result prediction was achieved using statistical tools, respectively. Another research reported the use of ^{1}HNMR for detection of 10%–50% adulteration of EVOO with hazelnut, peanut, soybean, and sunflower oil as major adulterants. The results were statistically analyzed using LDA and an excellent classification rate of 98 to 100% was observed [118]. A similar study was reported on camellia oil for discrimination of pure oil samples from samples admixed with sunflower, corn, or rapeseed oil. The ^{1}HNMR spectra were analyzed using PCA and PLS chemometrics and discrimination accuracy of nearly 90% was recorded [119].

Fluorescence spectroscopy is also reported to have been used for adulteration detection in EVOO by various researchers. Mabood *et al* used synchronous fluorescence spectroscopy to detect 2%–20% admixing of refined olive oil in VOO. The detection accuracy was observed to be high with a correlation coefficient of 0.94, obtained by applying PLSR statistical analysis [120]. A synonymous study was quoted on detection of pomace olive oil in EVOO adulterated in the level of 3%–33%. The synchronous fluorescence spectra were further analyzed by LDA and UPLS and high detection accuracy (99%) was recorded [121]. In another study, EVOO adulterated with 1%–30% of soybean oil was analyzed by fluorescence spectroscopy for detection of admixing. The researchers were able to detect adulteration, even at 1.33% of the adulterant, with a very high accuracy (R^2 0.98) [122]. Moreover, Li *et al* recently used PCA, LDA, and PLSR to analyze laser-induced fluorescence spectra to detect the presence of soybean oil in EVOO at 2%. This study also indicated that heat treatment applied before scanning the spectra greatly improved the efficacy of blending determination since R^2 of 0.92 and RMSEP of 0.27% were recorded [123].

11.3.6 Detection of lipid oxidation

Lipid oxidation is a series of complex reactions that cause the breakdown of oils and is one of the major causes of quality deterioration in edible oils. It is the fundamental unfavorable process that generates primary oxidation products (hydroperoxides), which then degrade into secondary oxidation products (such as aldehydes, ketones, acids, alcohols, lactones, and hydrocarbons)[124]. When heated, these secondary compounds become more stable and can have a detrimental effect on oil flavor, shelf life, and nutrition [125]. Moreover, many factors impact oxidation, including heat, light, degree of unsaturation, processing technique and the presence of antioxidants. Thus, oxidation stability of edible oils is assessed as a quality indicator throughout manufacturing, handling, and storage. Various tests, such as peroxide value and acidity, are conducted to determine the oxidation status in oils. If the PV of edible oils exceeds 3 meq kg^{-1}, then they are considered oxidized [25]. Additionally, K_{232} and K_{270} are considered as critical markers for good quality VOO because they indicate the formation of conjugated dienes, and conjugated trienes and carbonyl compounds, which are primary and secondary oxidation products, respectively, and thus are determined to assess the quality of oil [126].

Spectroscopic techniques have been used to estimate the oxidation stability of various edible oils (table 11.6). A study was conducted to investigate the use of

Table 11.6. A summary overview of various spectroscopic techniques used for estimation of oxidative stability of edible oils

Objective	Oil	Method	Multivariate analysis	Prediction statistics/results	References
Prediction of the peroxide value, K232 and K270, oil quality parameters	EVOO	NIR 11 500–4000 cm^{-1}	PLSR	Excellent prediction of k232 and k270 with R^2 of 0.9 and 0.9, respectively.	[31]
Determine the oxidative stability of oils	OO	FTIR 4000–650 cm^{-1}	PLSR	Correlation coefficient ranging between 0.87 and 0.99 was observed.	[33]
Determine the oxidative stability of oils	CO, PNO, SFO, SFFO, CAO, CSO	FTIR 4000–700 cm^{-1}	PLSR	Correlation coefficient ranging between 0.923 and 0.960 was observed.	[36]
Evaluate the oxidative stability of oils	CO, SFO, SBO, OO	UV 300–540 nm	MCR-ALS	An average % variance ranging between 87.15 and 98.95 was observed for the oil samples.	[49]
Detection of admixing of fried mustard oil in pure oil	MO	FTIR 1260–1080 cm^{-1}	PLSR	Correlation coefficient and root mean square error of 0.99 and 0.53% was observed.	[111]
Detection of admixing of fried coconut oil in pure oil	CNO	FTIR 1800–500 cm^{-1}	PLSR	Correlation coefficient and root mean square error of 0.997 and 0.832% was observed.	[127]
Prediction of the peroxide value, K232 and K270, oil quality parameters	OO	Raman 200–27 000 cm^{-1}	PCA and PLS	Key oxidation parameters were determined with high correlation and low root mean square error of 0.88, and 0.37% and 0.92 and 0.08% for K_{232} and for K_{270}.	[128]
Detection of lipid oxidation	CAO	SE-Raman 500–3000 cm^{-1}	PCA	SERS spectra showed significant differences between the initial CAO and the oil kept at 55°C for 5 days by PCA. In comparison to traditional measurements, the SERS technique was better able to identify changes in lipid molecules during the oxidation process.	[129]

Table 11.6. (*Continued*)

Objective	Oil	Method	Multivariate analysis	Prediction statistics/results	References
Detection of lipid oxidation	OO	Raman 2800–3100 cm^{-1} and 800–1800 cm^{-1}	-	RS was an effective tool for assessing oxidation in various grades of olive oil.	[130]
Analysis of oil quality parameters and prediction of its oxidation stability	EVOO, VOO	^{13}CNMR	SLRA	High accuracy of estimation was observed as 0.99991 correlation coefficient was recorded.	[131]
Prediction of oil oxidation stability	OO	^{1}HNMR	PCA	The findings of this investigation proved the great stability of olive oil. The PCA loadings, which demonstrate the contribution of each olive oil metabolite to its oxidative stability, revealed that 1H NMR signal intensities did not vanish or fluctuate significantly over the course of three and a half years.	[132]
Prediction of acidity in oil	EVOO	FS-detection wavelengths ranging from 197 nm to 998, 763 nm.	PLSR	High accuracy of detection was observed as correlation coefficient close to 0.9 was recorded.	[48]
Monitor the development of oxidation products at different oxidation durations	CMO	FS	PCA	—	[133]

BCO-black cumin oil; BO-blend oil; CAO-Canola oil; CMO-Camellia oil; CNO-Coconut oil; CO-corn oil; CSO-Cottonseed oil; EVOO-Extra VOO; FSO-Flaxseed oil; GO-Garlic oil; GSO-Grapeseed oil; HNO-hazelnut oil; LNO-Linseed oil; MO-Maize oil; MSO-Mustard oil; PNO-Peanut oil; PO-Palm oil; POO-Pomace olive oil; PSO-Pumpkin seed oil; RPO-Rapeseed oil; RO-Rice oil; ROO-refined olive oil ;ROPO-refined olive pomace oil; SAO-Sweet almond oil; SBO-Soybean oil; SFO-sunflower oil; SFFO-Safflower oil; SO-Soya oil; SSO-Sesame oil; VCNO-Virgin coconut oil; VO-Vegetable oil; WGO-wheat germ oil; WNO-Walnut oil.

UV–Vis spectroscopy as a feasible tool to compare the oxidative stability in corn, sunflower, soybean, and olive oil. The oil samples were heated from 30°C to 170°C, and the UV Vis spectra (300–540 nm) was compared for each sample heated at an interval of 10°C. Multivariate curve resolution with alternating least squares (MCR-ALS) and PCA chemometrics were employed to predict the increase in oxidation products on heating the oil. An average % variance ranging between 87.15 and 98.95 was observed for the oil samples [49]. Uncu and Ozen determined the oxidative stability of olive oil with Rancimat equipment, and the corresponding values were predicted from the FTIR spectra obtained at 40 000–650 cm^{-1} using the PLSR mathematical tool. A high prediction coefficient of 0.87–0.99 was obtained along with RMSEC (0.11) and RMSECV (0.86) values close to zero [33]. A synonymous study has been reported by Allendorf *et al* [36]. Another investigation on detection of admixing pure mustard oil and coconut oil with, respectively, fried and used oil has been reported. FTIR spectroscopy coupled with PLSR were employed to predict the admixing and a high correlation of 0.99 and 0.997, respectively, was obtained [111, 127].

In addition, spectroscopic techniques have been utilized for prediction of the oil quality indicators such as peroxide value, K_{232} and K_{270} in VOO. NIR (11 500–4000 cm^{-1}) spectroscopy coupled with PLSR is reported to have been able to generate excellent prediction with R^2 0.92, 0.9, and 0.9, for the aforementioned parameters, respectively [31]. A similar study was conducted on olive oil by employing low resolution Raman spectroscopy (200–2700 cm^{-1}) along with PCA and PLS. A total of 126 pure VOO and oxidized oils were tested for oxidative stability by studying their peroxide values, K_{232} and K_{270}. The oxidation indicators were determined quickly, in a non-destructive manner, and with correlation coefficients of 0.91, 0.88, and 0.90, respectively [128]. Another study reported the use of surface enhanced Raman spectroscopy (500–3000 cm^{-1}) and Raman spectroscopy with spectral range of 2800–3100 cm^{-1} to analyze the process of oxidation in canola oil and olive oil, respectively [129, 130].

NMR spectroscopy is reported to have been used to analyze oil quality parameters and predict its oxidation stability. A total of 66 oil samples comprising VOO from different cultivars corn, grapeseed, rapeseed, 'lampante' olive, refined olive pomace, refined olive, soybean, and sunflower oil were collected and subsequently ^{13}CNMR spectra were studied using stepwise linear regression analysis (SLRA) chemometrics. A high prediction correlation of 0.999 91 was observed [131]. A similar study was reported on olive oil, using ^{1}HNMR and PCA techniques [132]. In addition, Cao *et al* discovered that the fluorescence intensity (λ_{ex} = 320–700 nm) along with PCA could monitor the development of oxidation products in camellia oil, palm oil, perilla oil, and sunflower oil [133]. The acidity in EVOO was also estimated using fluorescence spectroscopy with wavelengths ranging between 197 and 998 nm, and with PLSR chemometrics. The prediction correlation coefficient was observed to be quite high (~0.9) [48].

11.4 Conclusion

Vegetable oils are key lipid sources in the human diet, with significant nutritional and calorific values. They are consumed and used all around the world, and therefore establishing their quality, authenticity, and adulteration is vital to human health. However, because of their comparable appearance, color, and chemical properties, there is always temptation to adulterate or mislabel premium edible oils for financial benefit. Despite widespread public awareness about the scale and consequences of oil adulteration, the incidence of fraud is difficult to assess. As a result, several analytical techniques for assessing the authenticity and quality of edible oils have been developed. In particular, the spectroscopic techniques serve as a quick, dependable, non-destructive, cost-effective, and real-time measurement approach for monitoring and assessment of oil quality. UV–Vis, IR, FTIR, Raman, NMR, and fluorescence spectroscopy are some of the spectroscopic techniques utilized in edible oil verification. Spectroscopic techniques are employed to estimate the physiochemical properties of oil; detect its bioactive components, geographical origin, and cultivar traceability; and to detect adulteration and lipid oxidation in various oil samples. Thus, this chapter addresses the numerous spectroscopic methods and their applications in the quality assessment of edible oils.

References

[1] Kumar A, Sharma A and C Upadhyaya K 2016 Vegetable oil: nutritional and industrial perspective *Curr. Genomics* **17** 230–40

[2] Rohman A, Ghazali M A, Windarsih A, Riyanto S, Yusof F M and Mustafa S 2020 Comprehensive review on application of FTIR spectroscopy coupled with chemometrics for authentication analysis of fats and oils in the food products *Molecules* **25** 5485

[3] Lohumi S, Lee S, Lee H and Cho B K 2015 A review of vibrational spectroscopic techniques for the detection of food authenticity and adulteration *Trends Food Sci. Technol.* **46** 85–98

[4] Ellis D I, Brewster V L, Dunn W B, Allwood J W, Golovanov A P and Goodacre R 2012 Fingerprinting food: current technologies for the detection of food adulteration and contamination *Chem. Soc. Rev.* **41** 5706–27

[5] Arvanitoyannis I S and Vlachos A 2007 Implementation of physicochemical and sensory analysis in conjunction with multivariate analysis towards assessing olive oil authentication/adulteration *Crit. Rev. Food Sci. Nutr.* **47** 441–98

[6] Valli E, Bendini A, Berardinelli A, Ragni L, Riccò B, Grossi M and Gallina Toschi T 2016 Rapid and innovative instrumental approaches for quality and authenticity of olive oils *Eur. J. Lipid Sci. Technol.* **118** 1601–19

[7] Zaroual H, El Hadrami E M and Karoui R 2021 Preliminary study on the potential application of Fourier-transform mid-infrared for the evaluation of overall quality and authenticity of Moroccan VOO *J. Sci. Food Agric.* **101** 2901–11

[8] Danezis G P, Tsagkaris A S, Camin F, Brusic V and Georgiou C A 2016 Food authentication: techniques, trends and emerging approaches *TrAC, Trends Anal. Chem.* **85** 123–32

[9] Penner M H 2017 Basic principles of spectroscopy *Food Analysis* pp 79–88

[10] Verma G and Mishra M 2018 Development and optimization of UV–Vis spectroscopy—a review *World J. Pharm. Res.* **11** 1170–80

[11] Clark B J, Frost T and Russell M A 1993 *UV Spectroscopy: Techniques, Instrumentation and Data Handling* (Berlin: Springer Science & Business Media)

[12] Kamal M and Karoui R 2015 Analytical methods coupled with chemometric tools for determining the authenticity and detecting the adulteration of dairy products: a review *Trends Food Sci. Technol.* **46** 27–48

[13] Casale M and Simonetti R 2014 Near infrared spectroscopy for analysing olive oils *J. Near Infrared Spectrosc.* **22** 59–80

[14] Stuart B 2000 *Infrared Spectroscopy* (New York: Wiley)

[15] Sun D W (ed) 2009 *Infrared Spectroscopy for Food Quality Analysis and Control* (New York: Academic)

[16] Rodriguez-Saona L E and Allendorf M E 2011 Use of FTIR for rapid authentication and detection of adulteration of food *Annu. Rev. Food Sci. Technol.* **2** 467–83

[17] van de Voort F R, Sedman J and Russin T 2001 Lipid analysis by vibrational spectroscopy *Eur. J. Lipid Sci. Technol.* **103** 815–26

[18] Rostron P, Gaber S and Gaber D 2016 Raman spectroscopy, review *Laser.* **21** 24

[19] Hu R, He T, Zhang Z, Yang Y and Liu M 2019 Safety analysis of edible oil products via Raman spectroscopy *Talanta* **191** 324–32

[20] Moro M K, Neto A C, Lacerda V, Romao W, Chinelatto L S, Castro E V and Filgueiras P R 2020 FTIR, 1H and 13C NMR data fusion to predict crude oils properties *Fuel* **263** 116721

[21] Zia K, Siddiqui T, Ali S, Farooq I, Zafar M S and Khurshid Z 2019 Nuclear magnetic resonance spectroscopy for medical and dental applications: a comprehensive review *Eur. J. Dent.* **13** 124–8

[22] Keeler J 2010 *Understanding NMR Spectroscopy* (New York: Wiley)

[23] Andueza D, Mourot B P, Aït-Kaddour A, Prache S and Mourot J 2015 Utilisation de la spectroscopie dans le procheinfrarouge et de la spectroscopie de fluorescence pourestimer la qualité et la traçabilité de la viande *INRA Prod. Anim.* **28** 197–208

[24] Zaroual H, Chénè C, El Hadrami E M and Karoui R 2022 Application of new emerging techniques in combination with classical methods for the determination of the quality and authenticity of olive oil: a review *Crit. Rev. Food Sci. Nutr.* **62** 4526–49

[25] Mousa M A, Wang Y, Antora S A, Al-Qurashi A D, Ibrahim O H, He H J, Liu S and Kamruzzaman M 2022 An overview of recent advances and applications of FT-IR spectroscopy for quality, authenticity, and adulteration detection in edible oils *Crit. Rev. Food Sci. Nutr.* **62** 8009–27

[26] Zhang W, Li N, Feng Y, Su S, Li T and Liang B 2015 A unique quantitative method of acid value of edible oils and studying the impact of heating on edible oils by UV–Vis spectrometry *Food Chem.* **185** 326–32

[27] Uncu O, Ozen B and Tokatli F 2019 Use of FTIR and UV–visible spectroscopy in determination of chemical characteristics of olive oils *Talanta* **201** 65–73

[28] Conceição J N, Marangoni B S, Michels F S, Oliveira I P, Passos W E, Trindade M A, Oliveira S L and Caires A R 2019 Evaluation of molecular spectroscopy for predicting oxidative degradation of biodiesel and vegetable oil: correlation analysis between acid value and UV–Vis absorbance and fluorescence *Fuel Process. Technol.* **183** 1–7

[29] Özdemir İ S, Dağ Ç, Makuc D, Ertaş E, Plavec J and Bekiroğlu S 2018 Characterisation of the Turkish and Slovenian extra VOOs by chemometric analysis of the presaturation 1H NMR spectra *LWT* **92** 10–5

[30] Cayuela-Sánchez J A, Palarea-Albaladejo J, García-Martín J F and del Carmen Perez-Camino M 2019 Olive oil nutritional labeling by using Vis/NIR spectroscopy and compositional statistical methods *Innovative Food Sci. Emerg. Technol.* **51** 139–47

[31] Willenberg I, Matthäus B and Gertz C 2019 A new statistical approach to describe the quality of extra VOOs using near infrared spectroscopy (NIR) and traditional analytical parameters *Eur. J. Lipid Sci. Technol.* **121** 1800361

[32] Garrido-Varo A, Sánchez M T, De la Haba M J, Torres I and Pérez-Marín D 2017 Fast, low-cost and non-destructive physico-chemical analysis of VOOs using near-infrared reflectance spectroscopy *Sensors* **17** 2642

[33] Uncu O and Ozen B 2015 Prediction of various chemical parameters of olive oils with Fourier transform infrared spectroscopy *LWT-Food Sci. Technol.* **63** 978–84

[34] Dogruer I, Uyar H H, Uncu O and Ozen B 2021 Prediction of chemical parameters and authentication of various cold pressed oils with fluorescence and mid-infrared spectroscopic methods *Food Chem.* **345** 128815

[35] Dupuy N, Galtier O, Ollivier D, Vanloot P and Artaud J 2010 Comparison between NIR, MIR, concatenated NIR and MIR analysis and hierarchical PLS model. Application to VOO analysis *Anal. Chim. Acta* **666** 23–31

[36] Allendorf M, Subramanian A and Rodriguez-Saona L 2012 Application of a handheld portable mid-infrared sensor for monitoring oil oxidative stability *J. Am. Oil Chem. Soc.* **89** 79–88

[37] Gurdeniz G, Ozen B and Tokatli F 2010 Comparison of fatty acid profiles and mid-infrared spectral data for classification of olive oils *Eur. J. Lipid Sci. Technol.* **112** 218–26

[38] Wójcicki K, Khmelinskii I, Sikorski M and Sikorska E 2015 Near and mid infrared spectroscopy and multivariate data analysis in studies of oxidation of edible oils *Food Chem.* **187** 416–23

[39] Liang P, Wang H, Chen C, Ge F, Liu D, Li S, Han B, Xiong X and Zhao S 2013 The use of Fourier transform infrared spectroscopy for quantification of adulteration in virgin walnut oil *J. Spectrosc.* **2013**

[40] Luna A S, da Silva A P, Ferré J and Boqué R 2013 Classification of edible oils and modeling of their physico-chemical properties by chemometric methods using mid-IR spectroscopy *Spectrochim. Acta, Part* A **100** 109–14

[41] Knothe G and Kenar J A 2004 Determination of the fatty acid profile by 1H-NMR spectroscopy *Eur. J. Lipid Sci. Technol.* **106** 88–96

[42] Hatzakis E, Dagounakis G, Agiomyrgianaki A and Dais P 2010 A facile NMR method for the quantification of total, free and esterified sterols in VOO *Food Chem.* **122** 346–52

[43] Dourou A M, Brizzolara S, Meoni G, Tenori L, Famiani F, Luchinat C and Tonutti P 2020 The inner temperature of the olives (cv. Leccino) before processing affects the volatile profile and the composition of the oil *Food Res. Int.* **129** 108861

[44] Guillén M D and Ruiz A 2003 Rapid simultaneous determination by proton NMR of unsaturation and composition of acyl groups in vegetable oils *Eur. J. Lipid Sci. Technol.* **105** 688–96

[45] Guillén M D and Uriarte P S 2009 Contribution to further understanding of the evolution of sunflower oil Unpublished to frying temperature in a domestic fryer: study by 1H nuclear magnetic resonance *J. Agric. Food Chem.* **57** 7790–9

[46] Zaroual H and Karoui R 2021 A preliminary study on the potential of front face fluorescence spectroscopy for the discrimination of Moroccan VOOs and the prediction of their quality *Anal. Methods* **13** 345–58

[47] Guzmán E, Baeten V, Pierna J A and García-Mesa J A 2015 Evaluation of the overall quality of olive oil using fluorescence spectroscopy *Food Chem.* **173** 927–34

[48] Baltazar P, Hernández-Sánchez N, Diezma B and Lleó L 2019 Development of rapid extra VOO quality assessment procedures based on spectroscopic techniques *Agronomy* **10** 41

[49] Gonçalves R P, Março P H and Valderrama P 2014 Thermal edible oil evaluation by UV–Vis spectroscopy and chemometrics *Food Chem.* **163** 83–6

[50] Feng S, Gao F, Chen Z, Grant E, Kitts D D, Wang S and Lu X 2013 Determination of α-tocopherol in vegetable oils using a molecularly imprinted polymers–surface-enhanced Raman spectroscopic biosensor *J. Agric. Food Chem.* **61** 10467–75

[51] Reddy L H and Couvreur P 2009 Squalene: a natural triterpene for use in disease management and therapy *Adv. Drug Deliv. Rev.* **61** 1412–26

[52] Baeten V, Dardenne P and Aparicio R 2001 Interpretation of Fourier transform Raman spectra of the unsaponifiable matter in a selection of edible oils *J. Agric. Food Chem.* **49** 5098–107

[53] Squeo G, Caponio F, Paradiso V M, Summo C, Pasqualone A, Khmelinskii I and Sikorska E 2019 Evaluation of total phenolic content in VOO using fluorescence excitation–emission spectroscopy coupled with chemometrics *J. Sci. Food Agric.* **99** 2513–20

[54] ISMEA 2020 *Esportare Olio Extra Vergine di Oliva negli Stati Uniti* (Rome, Italy: ISMEA)

[55] Baeten V and Aparicio R 2000 Edible oils and fats authentication by Fourier transform Raman spectrometry *BASE* **4** 196–203

[56] Mendez J, Mendoza L, Cruz-Tirado J P, Quevedo R and Siche R 2019 Trends in application of NIR and hyperspectral imaging for food authentication *Sci. Agropecu.* **10** 143–61

[57] Department of Agriculture 2010 *Agricultural Marketing Service United States Standards for Grades of Olive Oil and Olive-Pomace Oil* (Washington, DC: Federal Register; Department of Agriculture) 75

[58] Council E E C 1992 Council Regulation (EEC) No. 2081/92 on the protection of geographical indications and designations of origin for agricultural products and foodstuff *Off. J. Eur. Commun.*

[59] Dais P and Hatzakis E 2013 Quality assessment and authentication of VOO by NMR spectroscopy: a critical review *Anal. Chim. Acta* **765** 1–27

[60] Casale M, Oliveri P, Casolino C, Sinelli N, Zunin P, Armanino C, Forina M and Lanteri S 2012 Characterisation of PDO olive oil Chianti Classico by non-selective (UV–visible, NIR and MIR spectroscopy) and selective (fatty acid composition) analytical techniques *Anal. Chim. Acta* **712** 56–63

[61] Pizarro C, Rodríguez-Tecedor S, Pérez-Del-Notario N, Esteban-Díez I and González-Sáiz J M 2013 Classification of Spanish extra VOOs by data fusion of visible spectroscopic fingerprints and chemical descriptors *Food Chem.* **138** 915–22

[62] Lin P, Chen Y and He Y 2012 Identification of geographical origin of olive oil using visible and near-infrared spectroscopy technique combined with chemometrics *Food Bioprocess Technol.* **5** 235–42

[63] Devos O, Downey G and Duponchel L 2014 Simultaneous data pre-processing and SVM classification model selection based on a parallel genetic algorithm applied to spectroscopic data of olive oils *Food Chem.* **148** 124–30

[64] Tapp H S, Defernez M and Kemsley E K 2003 FTIR spectroscopy and multivariate analysis can distinguish the geographic origin of extra VOOs *J. Agric. Food Chem.* **51** 6110–5

[65] Portarena S, Baldacchini C and Brugnoli E 2017 Geographical discrimination of extra-VOOs from the Italian coasts by combining stable isotope data and carotenoid content within a multivariate analysis *Food Chem.* **215** 1–6

[66] Sánchez-López E, Sánchez-Rodríguez M I, Marinas A, Marinas J M, Urbano F J, Caridad J M and Moalem M 2016 Chemometric study of Andalusian extra VOOs Raman spectra: qualitative and quantitative information *Talanta* **156** 180–90

[67] Longobardi F, Ventrella A, Napoli C, Humpfer E, Schütz B, Schäfer H, Kontominas M G and Sacco A 2012 Classification of olive oils according to geographical origin by using 1H NMR fingerprinting combined with multivariate analysis *Food Chem.* **130** 177–83

[68] Lia F, Vella B, Zammit Mangion M and Farrugia C 2020 Application of 1H and 13C NMR fingerprinting as a tool for the authentication of Maltese extra VOO *Foods* **9** 689

[69] Winkelmann O and Küchler T 2019 Reliable classification of olive oil origin based on minor component profile using 1H-NMR and multivariate analysis *Eur. J. Lipid Sci. Technol.* **121** 1900027

[70] Calò F, Girelli C R, Angilè F, Del Coco L, Mazzi L, Barbini D and Fanizzi F P 2021 1H-NMR profiling shows as specific constituents strongly affect the international EVOO blends characteristics: the case of the Italian oil *Molecules* **26** 2233

[71] Jiménez-Carvelo A M, Lozano V A and Olivieri A C 2019 Comparative chemometric analysis of fluorescence and near infrared spectroscopies for authenticity confirmation and geographical origin of Argentinean extra VOOs *Food Control* **96** 22–8

[72] Lia F, Formosa J P, Zammit-Mangion M and Farrugia C 2020 The first identification of the uniqueness and authentication of Maltese extra VOO using 3D-fluorescence spectroscopy coupled with multi-way data analysis *Foods* **9** 498

[73] Al Riza D F, Kondo N, Rotich V K, Perone C and Giametta F 2021 Cultivar and geographical origin authentication of Italian extra VOO using front-face fluorescence spectroscopy and chemometrics *Food Control* **121** 107604

[74] Maestrello V, Solovyev P, Bontempo L, Mannina L and Camin F 2022 Nuclear magnetic resonance spectroscopy in extra VOO authentication *Compr. Rev. Food Sci. Food Saf.* **21** 4056–75

[75] Laroussi-Mezghani S, Vanloot P, Molinet J, Dupuy N, Hammami M, Grati-Kamoun N and Artaud J 2015 Authentication of Tunisian VOOs by chemometric analysis of fatty acid compositions and NIR spectra. Comparison with Maghrebian and French VOOs *Food Chem.* **173** 122–32

[76] Sinelli N, Casale M, Di Egidio V, Oliveri P, Bassi D, Tura D and Casiraghi E 2010 Varietal discrimination of extra VOOs by near and mid infrared spectroscopy *Food Res. Int.* **43** 2126–31

[77] Revelou P K, Pappa C, Kakouri E, Kanakis C D, Papadopoulos G K, Pappas C S and Tarantilis P A 2021 Discrimination of botanical origin of olive oil from selected Greek cultivars by SPME-GC-MS and ATR-FTIR spectroscopy combined with chemometrics *J. Sci. Food Agric.* **101** 2994–3002

[78] Shaw A D, di Camillo A, Vlahov G, Jones A, Bianchi G, Rowland J and Kell D B 1997 Discrimination of the variety and region of origin of extra VOOs using 13C NMR and multivariate calibration with variable reduction *Anal. Chim. Acta* **348** 357–74

[79] Jha S N, Jaiswal P, Grewal M K, Gupta M and Bhardwaj R 2016 Detection of adulterants and contaminants in liquid foods—a review *Crit. Rev. Food Sci. Nutr.* **56** 1662–84

[80] Jee M (ed) 2002 *Oils and Fats Authentication* (Boca Raton, FL: CRC Press)

[81] Azadmard-Damirchi S and Torbati M 2015 Adulterations in some edible oils and fats and their detection methods *J. Food Qual. Hazards Control* **2** 38–44

[82] Salah W A and Nofal M 2021 Review of some adulteration detection techniques of edible oils *J. Sci. Food Agric.* **101** 811–9
[83] Pooja B, Biswal A, Sarathchandra G and Selvam P Detection methods for assessing and quantifying the adulterants in edible cooking oil: a review
[84] Jee M 2009 Adulteration and authentication of oils and fats: an overview *Oils and Fats Authentication* **12** pp 1–24
[85] Navya P, Raju K and Sukumaran M K 2017 Analysis of food adulterants in selected food items purchased from local grocery stores *Int. J. Adv. Sci. Res.* **3** 82–9
[86] Das M and Khanna S K 1997 Clinicoepidemiological, toxicological, and safety evaluation studies on argemone oil *Crit. Rev. Toxicol.* **27** 273–97
[87] Yadav S 2018 Edible oil adulterations: current issues, detection techniques, and health hazards *IJCS* **6** 1393–7
[88] Aroca-Santos R, Lastra-Mejías M, Cancilla J C and Torrecilla J S 2019 Linear and non-linear quantification of extra VOO, soybean oil, and sweet almond oil in blends to assess their commercial labels *J. Food Compos. Anal.* **75** 70–4
[89] Torrecilla J S, Rojo E, Domínguez J C and Rodríguez F 2010 Linear and non linear chemometric models to quantify the adulteration of extra VOO *Talanta* **83** 404–9
[90] Jiang L, Zheng H and Lu H 2015 Application of UV spectrometry and chemometric models for detecting olive oil-vegetable oil blends adulteration *J. Food Sci. Technol.* **52** 479–85
[91] Yuan J J, Wang C Z, Chen H X, Ye J Z and Zhou H 2016 Identification and detection of adulterated Camellia Oleifera Abel. Oils by near infrared transmittance spectroscopy *Int. J. Food Prop.* **19** 300–13
[92] Chu X, Wang W, Li C, Zhao X and Jiang H 2018 Identifying camellia oil adulteration with selected vegetable oils by characteristic near-infrared spectral regions *J. Innov. Opt. Health Sci.* **11** 1850006
[93] Christy A A, Kasemsumran S, Du Y and Ozaki Y 2004 The detection and quantification of adulteration in olive oil by near-infrared spectroscopy and chemometrics *Anal. Sci.* **20** 935–40
[94] Vanstone N, Moore A, Martos P and Neethirajan S 2018 Detection of the adulteration of extra VOO by near-infrared spectroscopy and chemometric techniques *Food Qual. Saf.* **2** 189–98
[95] Jiang H and Chen Q 2019 Determination of adulteration content in extra VOO using FT-NIR spectroscopy combined with the BOSS–PLS algorithm *Molecules* **24** 2134
[96] Sohng W, Park Y, Jang D, Cha K, Jung Y M and Chung H 2020 Incorporation of two-dimensional correlation analysis into discriminant analysis as a potential tool for improving discrimination accuracy: near-infrared spectroscopic discrimination of adulterated olive oils *Talanta* **212** 120748
[97] Oussama A, Elabadi F, Platikanov S, Kzaiber F and Tauler R 2012 Detection of olive oil adulteration using FT-IR spectroscopy and PLS with variable importance of projection (VIP) scores *J. Am. Oil Chem. Soc.* **89** 1807–12
[98] Rohman A and Man Y C 2010 Fourier transform infrared (FTIR) spectroscopy for analysis of extra VOO adulterated with palm oil *Food Res. Int.* **43** 886–92
[99] Rohman A, Che Man Y B, Ismail A and Hashim P 2017 FTIR spectroscopy coupled with chemometrics of multivariate calibration and discriminant analysis for authentication of extra VOO *Int. J. Food Prop.* **20** S1173–81

[100] Sun X, Lin W, Li X, Shen Q and Luo H Y 2015 Analytical methods detection and quantification of extra VOO adulteration with edible oils by FT-IR spectroscopy and chemometrics *Anal Methods* **7** 3939–45

[101] Vasconcelos M, Coelho L, Barros A and de Almeida J M 2015 Study of adulteration of extra VOO with peanut oil using FTIR spectroscopy and chemometrics *Cogent Food Agric.* **1** 1018695

[102] Mendes T O, da Rocha R A, Porto B L, de Oliveira M A, dos Anjos V D and Bell M J 2015 Quantification of extra-VOO adulteration with soybean oil: a comparative study of NIR, MIR, and Raman spectroscopy associated with chemometric approaches *Food Anal. Methods* **8** 2339–46

[103] Rohman A and Che Man Y B 2012 Quantification and classification of corn and sunflower oils as adulterants in olive oil using chemometrics and FTIR spectra *Sci. World J.* **1** 250795

[104] Rohman A and Man Y B 2012 The chemometrics approach applied to FTIR spectral data for the analysis of rice bran oil in extra VOO *Chemometr. Intell. Lab. Syst.* **110** 129–34

[105] Rohman A and Che Man Y B 2012 Authentication of extra VOO from sesame oil using FTIR spectroscopy and gas chromatography *Int. J. Food Prop.* **15** 1309–892

[106] Maggio R M, Cerretani L, Chiavaro E, Kaufman T S and Bendini A 2010 A novel chemometric strategy for the estimation of extra VOO adulteration with edible oils *Food Control* **21** 890–5

[107] Chen H, Lin Z and Tan C 2018 Fast quantitative detection of sesame oil adulteration by near-infrared spectroscopy and chemometric models *Vib. Spectrosc.* **99** 178–83

[108] Rohman A and Man Y B 2011 The use of Fourier transform mid infrared (FT-MIR) spectroscopy for detection and quantification of adulteration in virgin coconut oil *Food Chem.* **129** 583–8

[109] Jamwal R, Kumari S, Dhaulaniya A S, Balan B, Kelly S, Cannavan A and Singh D K 2020 Utilizing ATR-FTIR spectroscopy combined with multivariate chemometric modelling for the swift detection of mustard oil adulteration in virgin coconut oil *Vib. Spectrosc.* **109** 103066

[110] Rohman A, Che Man Y and Ali M D 2019 The authentication of virgin coconut oil from grape seed oil and soybean oil using ftir spectroscopy and chemometrics *Int. J. Appl. Pharm.* **11** 259–63

[111] Jamwal R, Kumari S, Balan B, Dhaulaniya A S, Kelly S, Cannavan A and Singh D K 2020 Attenuated total reflectance–Fourier transform infrared (ATR–FTIR) spectroscopy coupled with chemometrics for rapid detection of argemone oil adulteration in mustard oil *LWT* **120** 108945

[112] El-Abassy R M, Donfack P and Materny A 2009 Visible Raman spectroscopy for the discrimination of olive oils from different vegetable oils and the detection of adulteration *J. Raman Spectrosc.* **40** 1284–9

[113] Becze A and Simedru D 2020 Rapid detection of walnut and pumpkin oil adulteration using Raman spectroscopy and partial least square methodology *Not. Bot. Horti Agrobot. Cluj-Napoca* **48** 1426–38

[114] Zhang X F, Zou M Q, Qi X H, Liu F, Zhang C and Yin F 2011 Quantitative detection of adulterated olive oil by Raman spectroscopy and chemometrics *J. Raman Spectrosc.* **42** 1784–8

[115] Georgouli K, Del Rincon J M and Koidis A 2017 Continuous statistical modelling for rapid detection of adulteration of extra VOO using mid infrared and Raman spectroscopic data *Food Chem.* **217** 735–42

[116] Zhu W, Wang X and Chen L 2017 Rapid detection of peanut oil adulteration using low-field nuclear magnetic resonance and chemometrics *Food Chem.* **216** 268–74

[117] Agiomyrgianaki A, Petrakis P V and Dais P 2010 Detection of refined olive oil adulteration with refined hazelnut oil by employing NMR spectroscopy and multivariate statistical analysis *Talanta* **80** 2165–71

[118] Šmejkalová D and Piccolo A 2010 High-power gradient diffusion NMR spectroscopy for the rapid assessment of extra-VOO adulteration *Food Chem.* **118** 153–8

[119] Shi T, Zhu M, Chen Y, Yan X, Chen Q, Wu X, Lin J and Xie M 2018 1H NMR combined with chemometrics for the rapid detection of adulteration in camellia oils *Food Chem.* **242** 308–15

[120] Mabood F, Boqué R, Folcarelli R, Busto O, Jabeen F, Al-Harrasi A and Hussain J 2016 The effect of thermal treatment on the enhancement of detection of adulteration in extra VOOs by synchronous fluorescence spectroscopy and chemometric analysis *Spectrochim. Acta, Part* A **161** 83–7

[121] Merás I D, Manzano J D, Rodríguez D A and de la Peña A M 2018 Detection and quantification of extra VOO adulteration by means of autofluorescence excitation-emission profiles combined with multi-way classification *Talanta* **178** 751–62

[122] Milanez K D, Nóbrega T C, Nascimento D S, Insausti M, Band B S and Pontes M J 2017 Multivariate modeling for detecting adulteration of extra VOO with soybean oil using fluorescence and UV–Vis spectroscopies: a preliminary approach *LWT-Food Sci. Technol.* **85** 9–15

[123] Li Y, Chen S, Chen H, Guo P, Li T and Xu Q 2020 Effect of thermal oxidation on detection of adulteration at low concentrations in extra VOO: study based on laser-induced fluorescence spectroscopy combined with KPCA–LDA *Food Chem.* **309** 125669

[124] Casal S, Malheiro R, Sendas A, Oliveira B P and Pereira J A 2010 Olive oil stability under deep-frying conditions *Food Chem. Toxicol.* **48** 2972–9

[125] Vlachos N, Skopelitis Y, Psaroudaki M, Konstantinidou V, Chatzilazarou A and Tegou E 2006 Applications of Fourier transform-infrared spectroscopy to edible oils *Anal. Chim. Acta* **573** 459–65

[126] Abenoza M, De Las Heras P, Benito M, Oria R and Sánchez-Gimeno A C 2016 Changes in the physicochemical and nutritional parameters of Picual and Arbequina olive oils during frying *J. Food Process. Preserv.* **40** 353–61

[127] Jamwal R, Kumari S, Kelly S, Cannavan A and Singh D K 2020 Rapid detection of pure coconut oil adulteration with fried coconut oil using ATR-FTIR spectroscopy coupled with multivariate regression modelling *LWT* **125** 109250

[128] Guzmán E, Baeten V, Pierna J A and García-Mesa J A 2011 Application of low-resolution Raman spectroscopy for the analysis of oxidized olive oil *Food Control* **22** 2036–40

[129] Li Y, Driver M, Decker E and He L 2014 Lipid and lipid oxidation analysis using surface enhanced Raman spectroscopy (SERS) coupled with silver dendrites *Food Res. Int.* **58** 1–6

[130] Carmona M Á, Lafont F, Jiménez-Sanchidrián C and Ruiz J R 2014 Raman spectroscopy study of edible oils and determination of the oxidative stability at frying temperatures *Eur. J. Lipid Sci. Technol.* **116** 1451–6

[131] Hidalgo F J, Gómez G, Navarro J L and Zamora R 2002 Oil stability prediction by high-resolution 13C nuclear magnetic resonance spectroscopy *J. Agric. Food Chem.* **50** 5825–31
[132] Alonso-Salces R M, Holland M V and Guillou C 2011 1H-NMR fingerprinting to evaluate the stability of olive oil *Food Control* **22** 2041–6
[133] Cao J, Li C, Liu R, Liu X R, Fan Y and Deng Z Y 2017 Combined application of fluorescence spectroscopy and chemometrics analysis in oxidative deterioration of edible oils *Food Anal. Methods* **10** 649–58

Chapter 12

Vibrational spectroscopy to determine food polyphenols

Deepika Umrao, Vivek Kumar, Abdul Rahman and Sumaiya Fatima

Dietary polyphenols positively affect human health by preventing degenerative diseases. These polyphenols are highly desired in food and other consumable plant products. The optimum amount of polyphenols in food and beverages defines the quality of these products; hence, an effective industrial polyphenol estimation system is required. Vibrational technologies have been used to identify the chemical structure of unknown organic and inorganic compounds for decades. Recent prediction models developed using chemometric methods and vibrational spectroscopic techniques have prompted the application of these analytical techniques in the food, beverage, pharmaceutical, and agricultural industries. Research is being conducted to develop analytical systems for qualitative and quantitative analysis of raw or processed food, plants, and other biological materials. Vibrational spectroscopy, which includes infrared spectroscopy (IRS) and Raman spectroscopy (RS), has recently emerged as a useful tool for the assessment of different food polyphenols. This chapter summarizes the types of vibrational spectroscopy, the importance of food polyphenols, and the application of vibrational spectroscopy in estimating dietary polyphenols in the food and beverage industries.

12.1 Introduction

Vibrational spectroscopy examines molecular oscillations, which provide structural data of chemical species under investigation and allows chemical compound identification, enabling qualitative analysis of the sample. In vibrational spectroscopy, the concentration of a chemical component in the analyte corresponds to the intensity of the signal, which enables the qualitative analysis of chemical compounds based on vibrational spectral data. The vibrational spectroscopy data for a multicomponent system is complex and requires additional support from chemometric algorithms for efficient analysis [1]. Vibrational spectroscopy offers various

doi:10.1088/978-0-7503-5573-5ch12 © IOP Publishing Ltd 2023

analytical methods, which can be employed depending on the sample type and state and the information to be attained. The key uses of vibrational spectroscopy in the food sector are connected to quantitative study of the major components present in foods using near-infrared (NIR) spectroscopy, mid-infrared (MIR) spectroscopy and RS. However, with recent technological advances in Fourier transform (FT) instrumentation, sampling, laser sources, and detectors, new avenues of application based on FT-IR and FT-Raman have also been investigated for quantitative and qualitative analysis to characterize food components and investigate their structure-function relationships. Furthermore, polyphenol compounds in food are of tremendous interest and have received increased attention in recent years due to their health benefits, e.g., several epidemiological studies have shown their positive effect in health enhancement by lowering the risk or symptoms of chronic illnesses. Most polyphenols have been linked with anti-cancer, antioxidant, anti-obesity, and anti-inflammatory properties, in addition to their prominent effect on sensory attributes of the food samples [2, 3]. Several attempts have been made to develop highly selective and sensitive methods to measure polyphenols, with extensive research on different extraction/separation techniques (e.g., SPE and SPME), as well as to enhance the chromatographic techniques (e.g., HPLC and GC-MS) being used. However, these reliable procedures are time consuming, labor intensive, costly, and complicated. Furthermore, the extraction procedure employed may impact the chemical composition of phenolic compounds usage. Therefore, techniques with little or minimal sample preparation, and cost-effective approaches are being studied to supplement or replace traditional analytical instruments. Vibrational spectroscopic techniques in combination with chemometric approaches might be employed as alternatives for quantitative study of diverse polyphenol compounds. Hence, vibrational spectroscopies have been applied as powerful tools to determine polyphenol compounds in food. Consequently, this chapter discusses various vibrational spectroscopy techniques, their advantages, and application for determination of polyphenols in food samples.

12.2 Vibrational spectroscopy

Vibrational spectroscopy is based on the transitions between quantized vibrational energy states of molecules. Two types of vibrational spectroscopy are commonly used for chemical compound analysis: RS and IRS, which includes near-infrared spectroscopy (NIRS) and mid-infrared spectroscopy (MIRS). Molecules in the sample vibrate and can change bond lengths or positions concerning other atoms in the molecule when exposed to radiation with suitable energy [4]; when the energy of molecular vibration falls in the infrared spectrum, the molecule can be characterized by IRS. The absorption of light with higher energy sometimes leads to the emission of radiation of different wavelengths, such as of incidence or in different directions (i.e. scattering), which leads to RS. The number of absorption signals, frequency, and intensity of absorption or emission provide an understanding of the different components of analytes. The number of absorption signals denotes the number of atoms and the shape of the compound. The frequency of the absorbed radiation denotes the energy of the bond and molecular vibrations, whereas the absorption intensity indicates analyte concentration [5]. The types of vibrational spectroscopy are explained in more detail in the following subsections.

12.2.1 Infrared spectroscopy

IRS is the most commonly employed spectroscopy technique, especially in biology and chemistry because of its applicability in the identification and structural analysis of chemical compounds [6], which leads to insight into the chemical and physical properties of compounds. IRS is sensitive to identifying different functional groups in a compound compared to UV and visible spectroscopy. It is based on atomic vibrations in a molecule. Compared with RS, the energy of an incident photon in IRS is entirely absorbed by the molecule and is not re-emitted, leading to molecular vibrations. An IR spectrum of the analyte is obtained by passing IR radiation through it and determining what fraction of IR radiation is absorbed by the molecules in the sample. Different bonds absorb different wavelengths of radiation depending on their strength, which allows the identification of functional groups or bonds in the specific wavelength. The frequency at which a peak is observed in the IR spectrum corresponds to the vibrational frequency of the molecule's fragment under investigation. A molecule or compound is identified by comparing its absorption peaks to a data bank of spectra. Each molecule has a characteristic spectrum or molecular fingerprint.

A characteristic feature of the molecule to show IR absorption is the net change in the molecule's dipole moment upon vibration (stretching or bending) or rotational motion of the molecule. An electromagnetic field is generated by an oscillating dipole moment upon molecular movement. This field interacts with the electromagnetic field accompanying incident radiation; absorption occurs whenever the frequency of incident radiation matches molecular pre-existing vibration or rotation, resulting in a change in the amplitude of molecular motion. IRS has been found to be helpful in qualitative and also quantitative analysis of multipart mixtures of similar organic or inorganic compounds. Molecular bond vibrations in the spectral range (in wavenumbers) of 12 500–10 cm^{-1} represent the infrared (IR) region. The NIR and MIR offer insight into the physical and chemical characteristics of the analyte, and have a spectral range of 12 500–4000 cm^{-1} and 4000–400 cm^{-1}, respectively. The far-infrared (FIR) spectral range falls in the zone of 400–10 cm^{-1}. Figure 12.1 depicts the working of IR spectroscopy. When the IR rays fall on the sample, upon their absorption the electrons of molecules can have electronic, vibrational, or rotational transitions depending upon the incident light wavelength. For rotational transitions, longer wavelengths are required, which is insufficient to cause vibrational or electronic transitions. Vibrational transitions can arise from the absorption of NIR light, and electronic transitions occur in the ultraviolet region. IR radiation absorption is characteristic of compounds with low energy differences between vibrational and rotational states [5].

The application of Fourier transformation to IRS, resulting in Fourier transformation infrared spectroscopy (FTIR), leads to a more advanced analytical

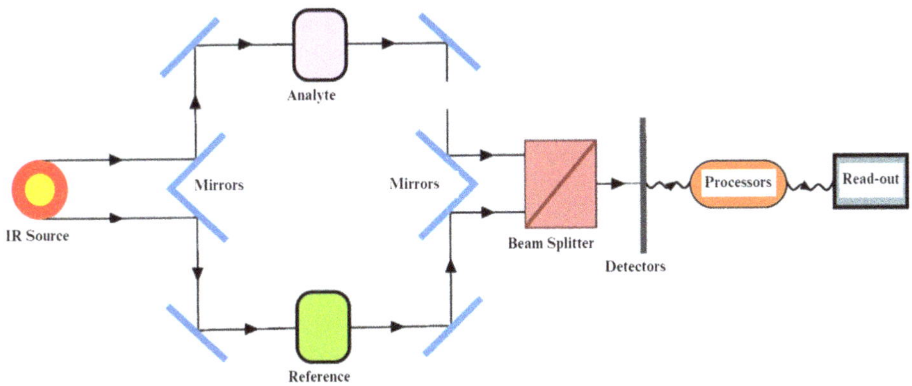

Figure 12.1. Schematic diagram of an infrared spectrometer.

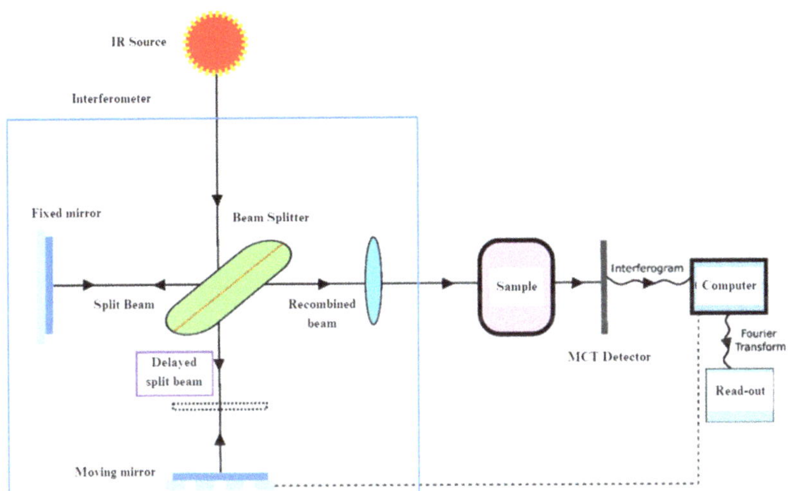

Figure 12.2. Schematic diagram of a Fourier transform-infrared spectrometer.

technique with short scanning time, higher resolution, and decreased background noise. FTIR employs a Michelson interferometer, which produces an interferogram, i.e., a signal produced as a function of variation in the path length of the two rays (figure 12.2). The mathematical method of the Fourier transformation interconverts domains of frequency and distance. An FTIR spectrometer usually comprises an IR radiation source, interferometer, sample cell, amplifier, detector, analog to digital (A/D) converter, and a computer. An FTIR spectrometer generates radiation and passes the sample through the interferometer. The detector detects the signal and sends it for amplification, removing contributions of high frequencies with a filter. The data is then converted into digital signals and displayed on the computer after a Fourier transformation. Tungsten-halogen lamps are used to produce NIR in FTIR. Nernst or Globar sources are used for the MIR region, whereas pressurized mercury lamps are used to produce FIR. The interferometer represents the core of FTIR spectrometers, which generates two beams of light from one incident light having a difference in path lengths. The interferometer then recombines these two beams and forwards them to the detector, which measures the intensity of these two beams as a function of path difference [5, 6].

One of the most commonly used interferometers is the Michelson interferometer, which consists of a beamsplitter and two perpendicular plane mirrors: a fixed and a movable mirror that moves in a direction perpendicular to the plane. The beamsplitter or semi-reflecting film is designed and placed to reflect half of the incident radiation to a movable mirror and transmit half of it to the fixed mirror. The two mirrors reflect these beams to the beamsplitter, which recombines them and then the interference can be observed. The transmitted beam from the interferometer perpendicular to the incident beam is detected in FTIR spectroscopy. If the difference between the fixed and stationary mirrors and beamsplitter is exact, then a zero-path difference (ZPD) will be observed. A path difference between two beams is observed as the movable mirror changes its position. The distance that the movable mirrors are displaced from the position of ZPD is defined as a mirror displacement (Δ), which makes the optical path difference (δ) equal to 2Δ. Path differences can be multiples of wavelength, or two beams can be shifted by half a wavelength, resulting in constructive and destructive interference, respectively. As the movable mirror moves, constructive, destructive, and interference between these two extremes are observed, resulting in increased or decreased signal intensity, which gives a cosine wave.

The Fourier transformation equation exists in pairs. One of the equations describes the intensity falling at the detector, $I(\delta)$. Which can be given by (Ng and Simmons 1999):

$$I(\delta) = \int_0^{\infty} P(\bar{\nu}) \cos 2\pi\bar{\nu}\delta.\, d\bar{\nu}. \tag{12.1}$$

where $\bar{\upsilon}$, δ and $P(\bar{\nu})$ are wavenumber, path difference, and spectral power density, respectively.

The other equation of the FT pair is given by:

$$P(\bar{\upsilon}) = \int_{-\infty}^{+\infty} I(\delta) \cos 2\pi\bar{\nu}\delta.\, d\delta. \tag{12.2}$$

These two equations are interconvertible by the Fourier transformation mathematical model. The first equation of the FT equation pair represents the variation in spectral power density as a function of path difference, whereas the second equation represents the variation in intensity as a function of wavenumber. The information collected by the detector generates an interferogram, which represents the relationship between response change over time and mirror scan distance. This time domain spectrum is further transformed to IR absorption spectrum, in which absorption intensity or percentage transmittance is plotted against wavenumber. Absorption (A) can be related to transmittance (T) by the following equation:

$$A = \log_{10}\frac{1}{T} = \log_{10}\frac{I_0}{I} \quad (12.3)$$

IRS has many advantages in the analytical field. This technique can provide information for the analytes at the whole range of the IR spectrum in a fraction of a second, which is helpful in the analysis of unstable compounds. IRS has a high resolution and higher sensitivity than a conventional spectrophotometer. In addition, IRS is a non-degradative to the sample and can analyze the pure sample in any phase, i.e., solid, liquid, or gas. However, samples containing water are not recommended for IRS because solvents sturdily absorb IR. Among IR spectroscopic techniques, FTIR spectroscopy has many advantageous over dispersive spectroscopic techniques. The two most significant advantages are multiplexity and the high throughput of FTIR spectroscopy. Multiplexity is due to enhancement in signal noise ratio per unit time, which is proportional to the square root of the number of resolution elements being measured. The high throughput advantage arises because slit devices are not used,0 which allows total IR radiation from the source to pass onto the sample continuously, leading to high energy falling on the detector and resulting in higher signals. Another advantage of FTIR spectroscopy is its high scanning speed: the moving mirror can move short distances rapidly. A key disadvantage of vibrational spectroscopic technique is the complex and hard-to-interpret IR spectrum, which makes identifying compound structure difficult using a single IR spectrum and requires further analysis using other spectroscopic methods, e.g., mass spectrometry (MS) and nuclear magnetic resonance (NMR), spectrum library, or a skilled operator to interpret the vibrational spectra of the compound [5].

12.2.2 Raman spectroscopy

RS is an advanced tool for identifying chemical species by analyzing radiation interaction with compounds. The molecular vibrational motion can be observed in RS if it changes the electron cloud polarizability of the molecule. When radiation in NIR and visible region falls on a molecule significant portion of it is absorbed and re-emitted in a different direction as incidence; this phenomenon is termed Rayleigh scattering. Here the molecule excited by the radiation reaches a virtual state and relaxes back to the original state, emitting photons of the same energy as of incident, resulting in elastic scattering. A small portion of the incident photon can transfer energy to the molecule (Stokes scattering) or take energy from it (anti-Stokes scattering), resulting in Raman scattering. In Stokes scattering, a molecule's electrons return to a higher vibrational state after being excited to a virtual state, emitting a photon with lower energy as of incident. In anti-Stokes scattering, the molecule is excited from its original vibrationally excited state to the exciting virtual state, and the electron relaxes back to the lower vibrational state, emitting a photon with higher energy and resulting in superelastic scattering. This change in the energy level of the scattered photon is measured by RS and can provide insight into the analyte concentration and chemical features [7].

The measurement is usually done using the change in wavenumbers from incident light radiation. For this reason, RS can use any wavelength radiation source, but

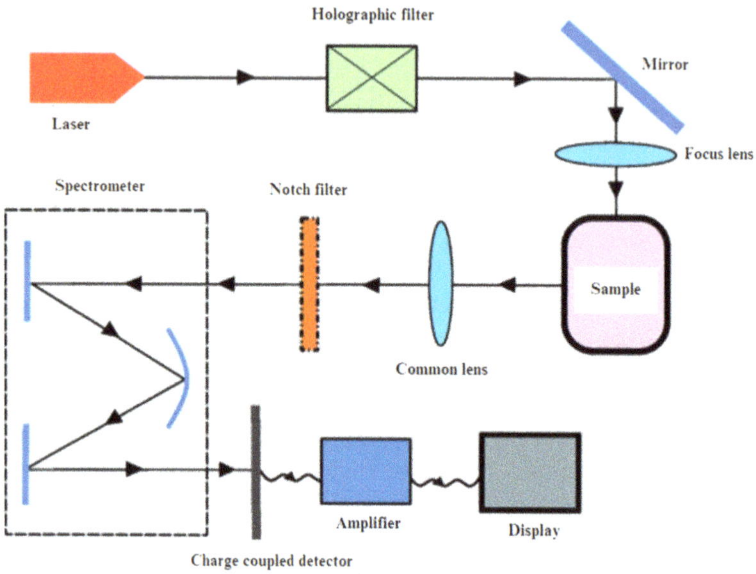

Figure 12.3. Schematic diagram of working of Raman spectrophotometer.

NIR and visible radiation sources are preferred. Photons from ultraviolet sources can cause photodecomposition of photosensitive samples and are rarely used. In RS, Stokes scattering is generally used because of the rare nature of anti-Stokes scattering, which requires electrons to be vibrationally excited before their transition and this weakens the intensity of the anti-stokes signal. RS measures the polarizability of the bond. Bonds with a high number of loosely held electrons have high polarizability and will have strong Raman spectroscopic signals. Therefore, RS provides more information about the molecular framework of the chemical species rather than the specific functional group, as in IRS. Figure 12.3 shows the working behind Raman spectroscopy.

Similar to IRS, RS can operate in dispersive and non-dispersive modes. Mercury arc lamps were used in earlier Raman spectrophotometers. In the late-1960s, laser sources become available, replacing Hg-arc lamps. Several varieties of laser sources have been employed in Raman spectrophotometers, e.g., NIR diode, Helium–Neon, Argon, and Krypton ion. A disadvantage of short-wavelength laser sources is their ability to produce fluorescence that photo-decomposes the sample. A laser coming from the source is collimated by a lens that sends it to an interference filter. Wavelength selectors or filters, such as double or triple-grating monochromators, rejection filters, or holographic filters, are used to separate weak Raman scattering signals from strong Rayleigh scattering signals. Photomultiplier tubes and photodiode array detectors were used in early RSs. However, technological and instrumental advances have replaced these detectors with charge transfer devices, such as charge-coupled and injection devices [7].

Although RS is a powerful tool for the characterization of unknown chemical species, the Raman signal is intrinsically frail and makes up only 0.001% of the total

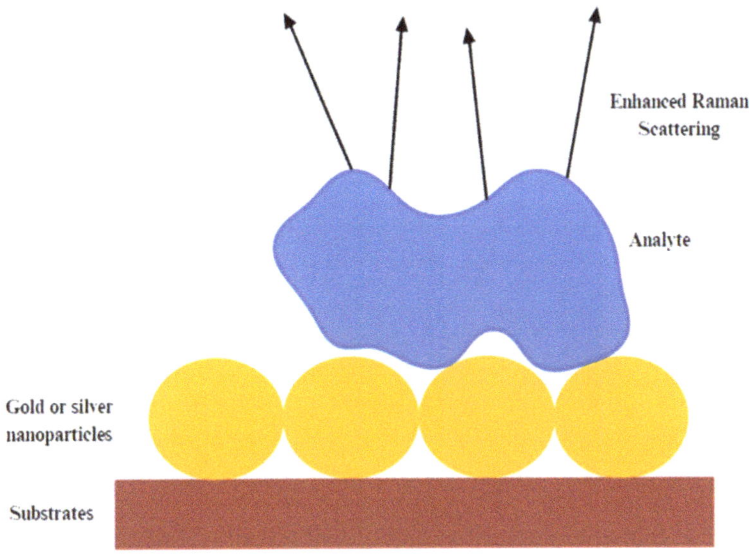

Figure 12.4. Main components of the surface enhanced Raman scattering system.

source radiation intensity. The sample is placed close to a metal surface in surface enhanced Raman spectroscopy (SERS), enhancing the weak Raman signal, which is achieved by increasing the electromagnetic field generated by the excitation of localized surface plasmons (LSPs) of molecules under investigation placed on the roughened metal surface of Gold/Silver nanoparticles (figure 12.4) [8]. SERS is a sensitive analytical technique that can detect small amounts of samples because of up to 15 × amplification in Raman signal compared with usual Raman scattering [9]. SERS has also been helpful in nanotechnology, which helps us to understand the chemistry occurring on a nanoparticle's surface.

RS can be used to identify both organic and inorganic compounds. Metals in coordination compounds have loosely bounded electrons, and therefore have intense Raman signals. These same compounds have low absorption frequencies in IRS [10]. The laser source used in RS can cause fluorescence of the analyte's organic component, resulting in decreased signal-to-noise ratio; therefore, RS data sometimes should be interpreted with other spectrophotometric data, such as IRS [11].

12.3 Vibrational spectroscopy in food analysis

Vibrational spectroscopy-based techniques are widely used in food industries for quality control. They deliver the wavelength of light that is absorbed by the molecule. The absorption of its spectrum is then compared with that of a known compound to identify an unknown compound. Vibrational spectroscopy can also be used to understand the physical features of chemical compounds. This technique is a better substitute to determine chemically diverse compounds in food samples than traditional methods, which apply toxic agents and require complex sample pretreatment. Vibrational spectroscopy can quickly analyze several chemical components

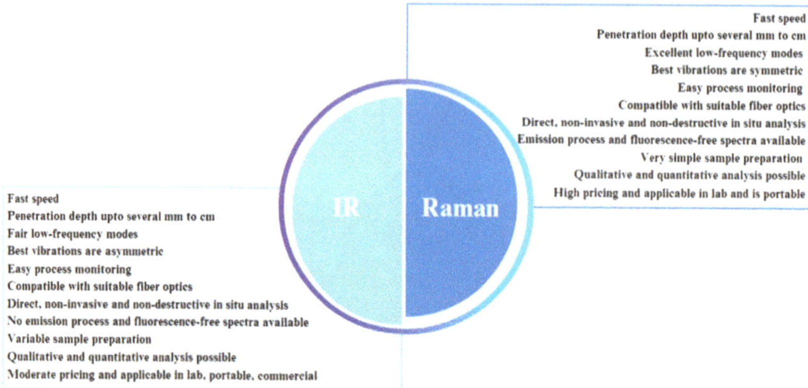

Figure 12.5. The significant features of vibrational spectroscopy to detect phenolic compounds.

from a single measurement. Furthermore, development in instrumentation and computational techniques has made vibrational techniques available for *in situ* and online measurement of various chemical components of food samples. Therefore, vibrational spectroscopy—indulging different spectroscopic techniques, e.g., FTIR, NIR, MIR, and Raman—when combined with the chemometric process could be employed for quantitative examination of numerous functional and aroma compounds in food commodities due to various significant advantages. Their significant features are depicted in figure 12.5 [12]. Vibrational spectroscopy has been utilized for quantitative determination of major food components, such as moisture, fat, protein, carbohydrate, along with the minor components, such as volatile compounds, vitamins, phenolic compounds, terepenoids, alkaloids, etc, in various food components [13–15].

12.4 Determination of polyphenols by vibrational spectroscopy

Dietary polyphenols are the derivative or isomers of flavonols/phenolic acids/isoflavones/flavones, which are metabolic products originating from the shikimate and phenylpropanoid pathway. Over 8000 structural variants of polyphenols are present in the form of plants' secondary metabolites, mainly consisting of aromatic rings with more than one hydroxyl group moieties, and are widely distributed in fruits, vegetables, tea, wine, extra virgin oil, chocolate etc. The experimental results support the role of dietary polyphenols to activate endogenous defense systems by regulating cellular signaling processes, e.g., N9-KB activation, glutathione metabolism process, AP-I DNA Binding, PI3-Kinase/MAPK protein pathway activation, and the translocation process into the nucleus. Additionally, the intake of polyphenols support vigorous antioxidant activity against reactive oxygen species (ROS), which are generated as result of oxidative stress and thus regulates the cell [16, 17]. Polyphenols also have an important role in the sensory attributes and bacteriological consequences of wines. Anthocyanidins, catechins, flavonol glycosides, flavonols, and tannins profoundly impact the flavor of wine [18]. Furthermore, fruit phenolic compounds, such as flavonoids, quercitin, revesrtol, etc, are linked to physiological activity, which is dependent on antioxidant characteristics, the capacity to prevent nitrosation, and the ability to chelate metal ions [19]. Dietary polyphenols are divided into subclasses based on their phenolic rings and the structural elements responsible for linking these aromatic rings, as shown in table 12.1 [20].

Phenolic acids are non-flavonoid polyphenolic chemicals that may be further classified into two types: benzoic acid derivatives and cinnamic acid derivatives with C1–C6 and C3–C6 backbones, respectively (table 12.1). While fruits and vegetables contain numerous free phenolic acids, phenolic acids are frequently bound in grains and seeds, notably in the bran or hull [21, 22]. Only acid or alkaline hydrolysis, or enzymes, can liberate or hydrolyze these phenolic acids. Furthermore, flavonoids have a C6–C3–C6 general structural backbone, with two phenolic C6 units (Ring A and Ring B) (table 12.1). Flavonoids are further classified as anthocyanins, flavonones, flavonols, flavon-3-ols, and flavones based on their hydroxylation pattern and differences in the chromane ring (Ring C). In addition, most of the flavonoids have their Ring B coupled to the C2 position of Ring C, while other flavonoids, such as isoflavones and neoflavonoids, have their Ring B attached to the C3 and C4 positions of Ring C, respectively. Despite the absence of the heterocyclic Ring C, chalcones are classified as flavonoid compounds. Flavonoids' fundamental structures are aglycones; nevertheless, in plants, the majority of these chemicals occur as glycosides. These molecules' biological properties, including antioxidant activity, are affected by both structural differences and glycosylation variations [23–33].

Functional substituents such as N may also be present in some polyphenols, being referred to as polyphenolic amides. The following two categories of flavonoids are significant because they are the principal constituents of common foods: capsaicinoids and avenanthramides found in chilli peppers [34] and oats [35] (table 12.1). The former, polyphenolic amide, such as capsaicin, are responsible for the heat in chilli peppers, impart powerful anti-inflammatory and antioxidant activities, and modulate the oxidative defense mechanism in cells. Moreover, the latter also has antioxidant properties, including the prevention of LDL oxidation. Additionally, apart from phenolic acids, flavonoids, and polyphenolic amides, non-flavonoid polyphenols are also found in foods, and are regarded to be helpful to human health. Revestrol is only found in grapes and red wine. Ellagic acid and its derivatives are found in berry fruits, such as strawberries and raspberries, as well as in the skins of some tree nuts, which are common examples of non-flavonoid polyphenols [36, 37]. Turmeric and ginger contain curcumin and gingerol, both being powerful antioxidants [38, 39]. Rosmarinic acid is a caffeic acid dimer, while ellagic acid is a gallic acid dimer. While gallic acid and ellagic acid are present in their free forms, their glucose esters, known ashydrolysabletannins, are also found in many plants [40] (table 12.1). Vibrational spectroscopy has been used to identify molecules and biological and aromatic compounds in food attributes. Raman, NIR, MIR and FT-IR successfully contribute to identifying various total phenolic, flavonoids, and anthocyanins content in different food items, which have a medicinal impact on health, as listed in table 12.2.

Table 12.1. Classification of phenolic compounds.

Classification	Dietary polyphenols	Chemical structure	Fruit sources	References
Phenolic acids	Benzoic acid and its derivatives	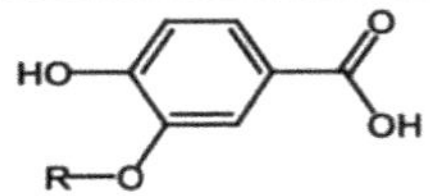	Apples, cranberries, prunes, strawberries, and yogurts	[21]
	Cinnamic acid and its derivatives	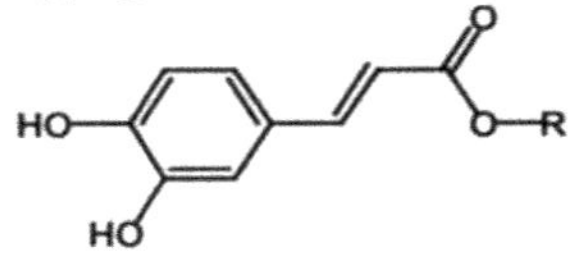	Brassicas vegetables, celery, citrus fruits, Cocoa, grapes, spinach, and tea	[21, 22]
Flavonoids	Isoflavones, neoflavonoids and chalcones	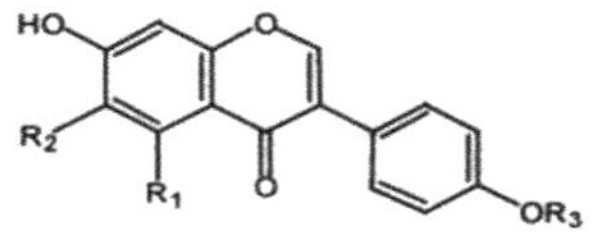 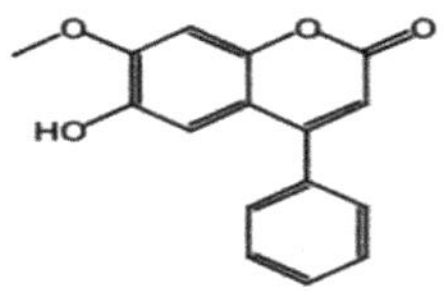 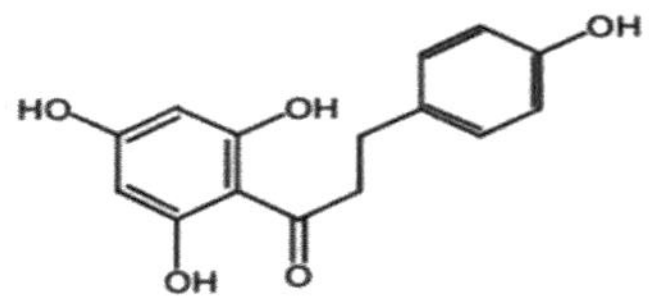	Mostly found in the leguminous plants; apples, and hops or beers	[23–28]
	Flavones, flavonols, flavonones and flavononols	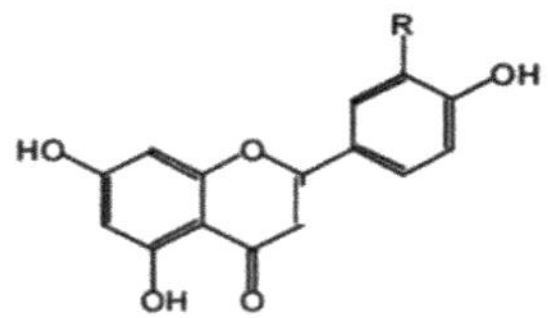	Apples, berries, cherries, citrus fruits, leafy vegetables, onions, soybeans, tea, and wine	[29–31]

(*Continued*)

Table 12.1. (*Continued*)

Classification	Dietary polyphenols	Chemical structure	Fruit sources	References
		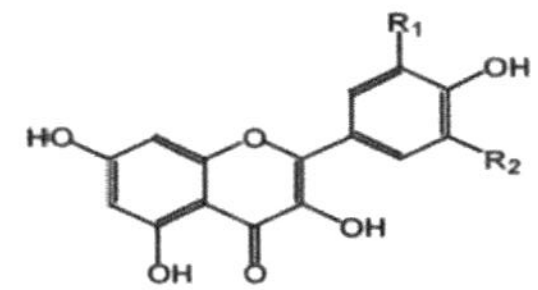		
		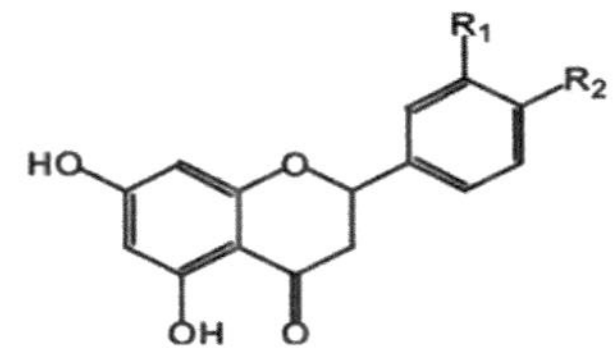		
		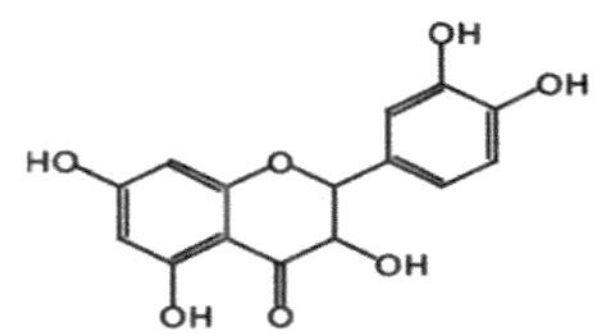		
	flavonols and proanthocyanidins	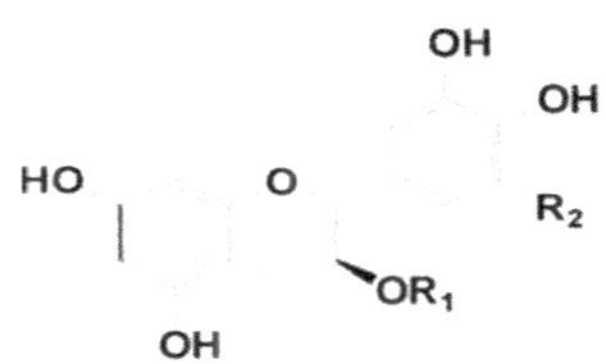	Found in many fruits, particularly in the skins of apples, blueberries and grapes	[27, 32, 33]

		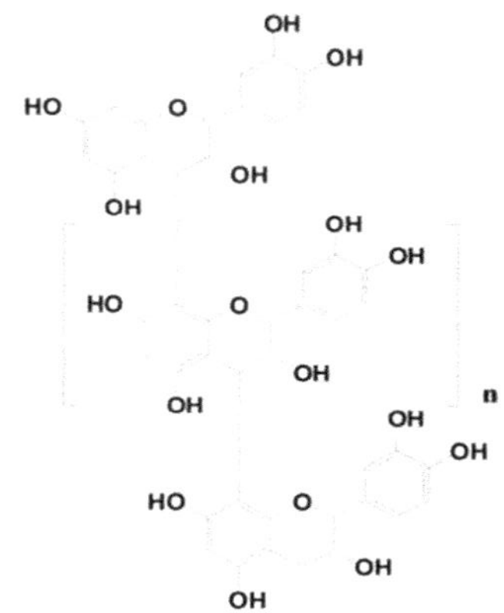		
	Anthocyanidins	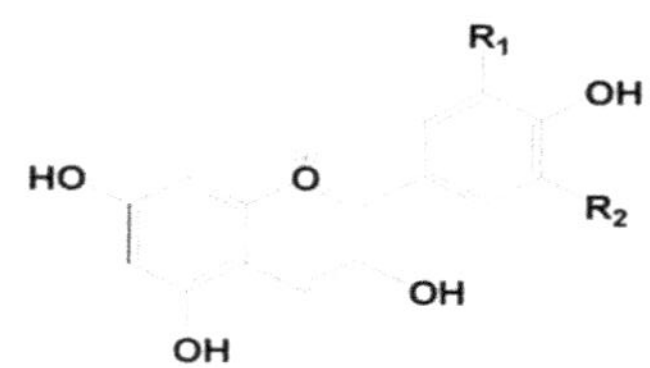	Red, purple, and blue fruits and vegetables, such as berries, plum, red wine, etc.	[30]
Polyphenolic amides	Capsaicinoids	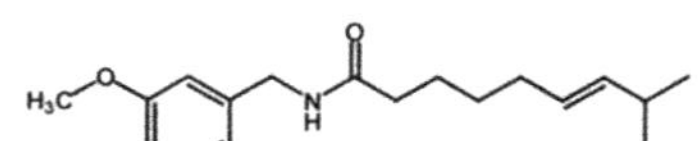 Capsaicin 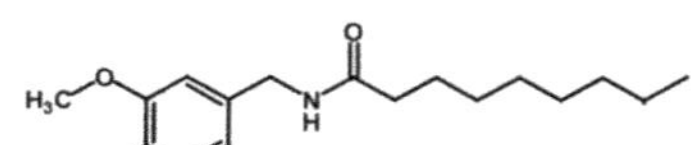 Dihydrocapsaicin	Chilli peppers	[34]
	Avenanthramides	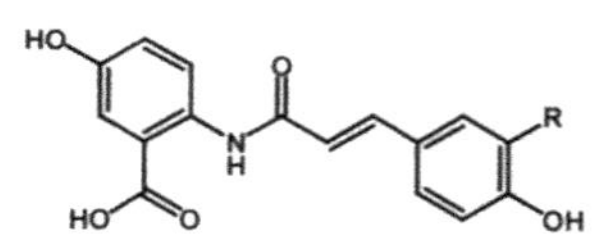	Oats	[35]

(*Continued*)

Table 12.1. *(Continued)*

Classification	Dietary polyphenols	Chemical structure	Fruit sources	References
Other polyp henols	Resveratrol	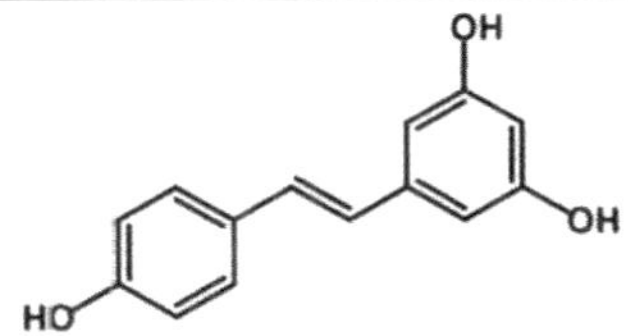	Grapes and red wine	[36]
	Ellagic acid and its derivatives	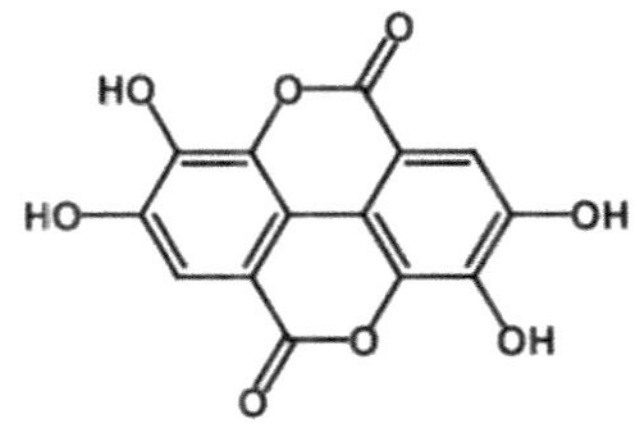	Berry fruits, e.g., strawberries and raspberries, and in the skins of different tree nuts	[37]
	Curcumin	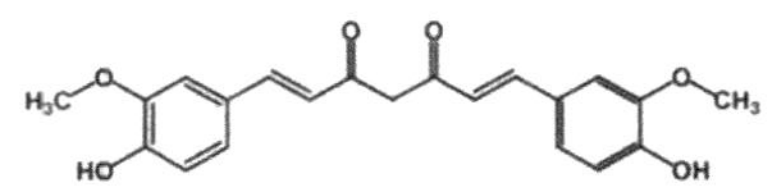	Turmeric	[38]
	Gingerol	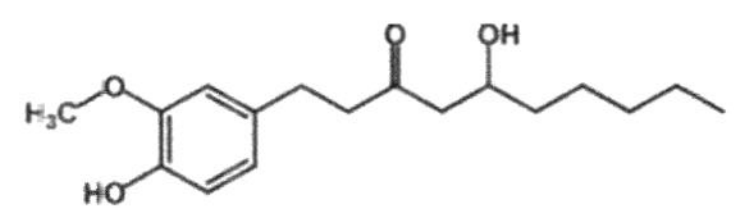	Ginger	[39]
	Rosmarinic acid	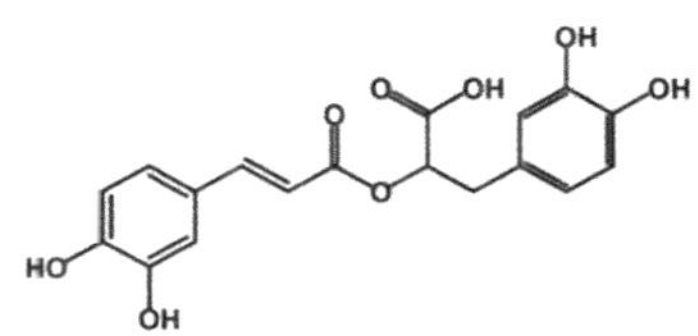	Lemon balm, oregano, peppermint, rosemary, sage, thyme	[40]

Table 12.2. Application of vibrational spectroscopy to detect phenolic compounds in various food products.

Food product	Phenolic compounds	Quantification method	Wavelength range	Chemometrics	Reference
NIR spectroscopy					
Olive oil	Alpha-beta-gamma tocopherols	HPLC-NIR	350–1000 nm	PLS-DA	[41]
Raw propolis	TFC and TPC	UV–NIR	1100–2000 nm	MPLS	[42]
Hawthorn fruit	TFC and TPC	UV–VIS–NIR	800–2500 nm	PCA, PLS	[43]
Apple	Ascorbic acid and TPC	HPLC–UV–NIR	400–2500 nm	LS-SVM	[44]
Dehydrated tomato	Lycopene and TPC	UV–VIS–NIR	800–2500 nm	RBF-NN	[45]
Guava nectar	Ascorbic acid	Titrimetric NIR	10 000 to 4000 cm^{-1}	MPLS	[46]
Cashew fruit	Ascorbic acid	Titrimetric NIR	10 000 to 4000 cm^{-1}	MPLS	[46]
Red grape skin spectrophotometer	TPC, TFC, Anthocyanins 908–1676 nm	Portable NIR	PLS	[47]	
Red wine	Quercetin, Catechin, Malvin, Oenin	HPLC–NIR	190–2500 nm	PCA, PLS	[48]
Tea powder, leaf, canopy	Polyphenols	NIRS	350–2500 nm	PLS	[52]
Green tea	TAC	HPLC–NIR	1100–2500 nm	PLS, SNV	[53]
Green tea	Catechins	FT-NIR	10 000–4000 cm^{-1}	SNV	[54]
Tea	TPC	NIR	900–1701 nm	PLS	[55]
sweet basil	Polyphenol, eugenol and terpene content	NIR	—	PLS-DA	[56]
Tea	Polyphenols	GC–MS–NIR	—	SNV	[57]
Tea	–	UV–VIS–NIR	350–1075 nm	ANN	[58]
Roasted coffee	Chlorogenic acid	HPLC–UV–NIR	1000–2700 nm	PLS	[59]
Honey	TFC and TPC	UV–NIR	10 000–4000 cm^{-1}	—	[60]
Quinoa	TPC	UV–VIS–NIR	1100–2500 nm	—	[64]
MIR spectroscopy					
Pomegranate juice	TPC, TAC, ascorbic acid	UV–VIS–MIR	5000–930 nm	PLS	[65]
Wine	Anthocyanin	MIR	4000–650 cm^{-1}	PLS	[66]
Australian Syrah wine	Total anthocyanins	ATR-MIR spectroscopy	1000 and 1100 cm^{-1}	PLS	[67]
Wine	Total phenolic and anthocyanin contents.	MIR	4000–650 cm^{-1}	PLS	[68]

(*Continued*)

Table 12.2. (*Continued*)

Food product	Phenolic compounds	Quantification method	Wavelength range	Chemometrics	Reference
FTIR spectroscopy					
Seagrass	Gallic acid, vanillin, tannic acid, and p-coumaric acid	FT-MIR	4000–400 cm^{-1}	—	[70]
Chickpea	TPC	UV–FTIR	4000–450 cm^{-1}	—	[71]
Okra seeds	TPC, Isoquercitrin, Quercetin-3-O-gentiobioside	HPLC–UV–FT-MIR	2000–4000 cm^{-1}	PLS	[72]
Apple	TPC, TFC	UV–VIS–FT-NIR	12 500–3600 cm^{-1}	PLS	[73]
Grapefruit	TAC and TPC	UV–VIS–FT-NIR	4000–400 cm^{-1}	—	[74]
Grape juice	TAC, TPC	UV–VIS–FT-MIR	4000–400 cm^{-1}	PLS	[75]
Pomegranate fruit	TPC, TAC, Ascorbic acid	UV–VIS–FT-NIR	1064–1333 nm	PLS	[76]
Honey	Catechin, Chlorogenic acid, Syringic acid, Vanillic acid	HPLC–DAD FT-IR	4000–600 cm^{-1}	PLS	[12]
Chocolate	(−)- Epicatechin, (+)-Catechin; TPC	UV–VIS–FT-IR; HPLC–DAD FT-IR	1800–700 cm^{-1}	PLS	[77]
Green tea	Epigallocatechin gallate, epigallocatechin, epicatechin gallate and epicatechin	NIR	4000–8500 cm^{-1}	PLS	[78]
Spent coffee	Caffeic acid, (+)-Catechin, Chlorogenic acid	HPLC–PDA FT-NIR	10 000–4000 cm^{-1}	PLS	[79]
Coffee beans	3-Caffeoylquinic acid, 4-Caffeoylquinic acid, 5-Caffeoylquinic acid, 3,5-Dicaffeoylquinic acid, Dicaffeoylquinic acid, 4,5-Dicaffeoylquinic acid	HPLC–DAD FT-IR	4000–400 cm^{-1}	PLS	[80]
Olive oil	Hydroxytyrosol and its derivatives, Tyrosol and its derivates, 3,4-DHPEA-EDA, o-Diphenols, p-HPEA-EA, Total phenols	HPLC–DAD FT-NIR	12 500–4000 cm^{-1}	PLS	[3]

Wine	Total phenolic compounds, total anthocyanins and total condensed tannins	FT-MIR	797–2989 cm^{-1}	PLS	[81]
Wine	Anthocyanin	FT-MIR	5012–926 cm^{-1}	PLS	[82]
Red wine	Tannin	FT-MIR transmission Spectroscopy	400–4000 cm^{-1}	SIMCA	[83]
Raman spectroscopy					
Bread	Total extractable polyphenol contents	Raman	200–3200 cm^{-1}	—	[84]
Potato chips	TPC	Raman	—	PCA	[85]
Potentilla tormentilla, herb	Total polyphenols, tannins, ellagitannins, procyanidins, agrimoniin, 3-O-galloylquininc acid and catechin	Raman	100–3700 cm^{-1}	PLS	[86]
Cistus incanus leaves	TPC, TFC, anthocyanins	Raman	100–3700 cm^{-1}	PCA	[87]
Plant gallnuts	Tannin, gallate, pyrogallol, and syringic acid	UV–Vis–Raman	100–2000 cm^{-1}	–	[89]
Red wine	Total polyphenols, total anthocyanin and total tannin contents	Portable Raman spectrometer	0–500 cm^{-1}	PLS	[90]
Honey	Catechin, chlorogenic acid, syringic acid, vanillic acid	HPLC-DAD Raman	50 to 3500 cm^{-1}	PLS	[12]
Carrots	Carotenoids, α-carotene, β-carotene, lutein, polyacetylenes, falcarinol, falcarindiol, 3-o-acetylfalcarindiol	HPLC-DAD FT-Raman	400–2500 nm	PLS	[91]

TPC: Total phenolic content; TFC: Total Flavonoid content; TAC: Total anthocyanins content; MPLS: Multiprotocol Label Switching; PCA:Principal component analysis; PLS:Partial least squares; PLS-DA:Partial least squares-Discriminant Analysis; SIMCA:Soft independent modeling by class analogy; SNV: Support Vector Machines;LS-SVM:Least squares support vector machines ;RBF-NN: Radial Basis Function Neural Network.

12.4.1 NIR spectroscopy to determine polyphenols

NIR spectroscopy has been widely applied in the determination of polyphenols in various food commodities. Different food items are enriched with a variety of polyphenols. Olive oil contains α-, β-, γ-tocopherols compounds that are reported to have been detected by the HPLC-NIR technique with 350–1000 cm^{-1} wavelength. These tocopherols are responsible for inhibiting lipid peroxidation, work as antioxidants, and have anti-aging properties. The (Rp) prediction value came up to 0.91 using HPLC combined with NIR, thus confirming its applicability [41]. In the case of determination of flavonoids and phenols in raw propolis, a glue extracted from bees and formulated in the form of tincture with significant impact on boosting the immune system. NIR reflectance with a wavelength of 1100–2000 nm is reported to have been used for detection of the total flavonoids and phenolic compound present [42].

NIRS is an advanced technology for qualitative and quantitative analysis. Total phenolic and flavonoid content along with some specific phenolic compounds have been determined in fruits and their products. In hawthorn fruit, Dong, Ni, and Kokot employed UV–Vis–NIR (800–2500 nm) to determine total phenolic and flavonoid content, and application of ANN, SVM and PLS models showed high correlation of calibration and prediction [43]. In another study, ascorbic acid and total phenolic content by employing HPLC–UV–NIR spectroscopy were measured in the spectral range of 400–2500 nm and LS-SVM [44]. In addition, lycopene content and TPC have been measured in dehydrated tomato by UV–VIS–NIR spectroscopy (800–2500 nm) and RBF-NN chemometrics [45]. In another study, ascorbic acid content has been measured in guava nectar and cashew fruit using titrimetric NIR (10 000–4000 cm^{-1}) and MPLS chemometrics [46]. Furthermore, anthocyanin, flavonoid, and phenolic content in the skin of red grapes have been measured using portable NIR spectrophotometer (908–1676 nm) and PLS statistical tools [47], while Quercetin, Catechin, Malvin, Oenin measured content in red wine using HPLC-NIR in the spectral range 190–2500 nm [48].

NIRS provides a quick and easy method to regulate valid tea constituents to administer industrial manufacturing. Tea has antioxidant properties because of its polyphenolic content (180–360 mg g^{-1}) and may prevent human health-related disorders [49]. Tea quality, texture, astringency, and color depend upon the concentration and composition of polyphenolic compounds [50]. Tea quality is determined by its biochemical composition, e.g., total tea polyphenols, amino acids, and soluble carbohydrates. Polyphenols are the main components of water-soluble pigments of tea. The main components of tea polyphenol are catechins and caffeine. Polyphenolic content in tea leaves decreases with aging—fresher leaves have more catechins and caffeine levels [51]. Therefore, the determination of polyphenolic content in tea leaves can be used to examine the degree of tea freshness. A study of *Camellia sinensis*, a variety of tea, has estimated the concentration of these chemicals by NIR reflectance spectroscopy in three different states; powder, leaf, and canopy [52]. NIR has been used to build a calibration model for the qualitative and quantitative analysis of *C. sinensis* by predicting the content of caffeine, epigallocatechin (EGCG), epicatechin (EC), and total antioxidant capacity [53]. The correlation coefficient (R) achieved was 0.93 for caffeine, 0.83 for EGCG, and 044 for EC. The applicability of a multivariate calibration model to NIR spectra to determine total polyphenols and caffeine content in green tea has been studied using the PLS algorithm to calibrate and achieve a better correlation coefficient of the prediction model, $R = 0.9688$ for caffeine and $R = 0.9299$ for total polyphenols [54].

An integrated quality examination system for tea leaves was developed for *in situ* polyphenol estimation using NIR reflectance spectroscopy [55]. This system was designed to carry out sample preparation, enzyme deactivation, drying, grinding and sieving, and polyphenol content estimation. The harvest time for six different cultivars of sweet basil was optimized by detecting biochemicals such as rosmarinic acid, eugenol, and terpene. Biochemical characterization of cultivars was done by combining NIRS, liquid and gas chromatography, and chemometric statistical methods of PLS [56]. A recent study has used nano-optical sensors with NIRS to rapidly estimate polyphenols [57]. The analytical technique was then used to investigate the *in situ* polyphenol consumption rate during the ultrasonic fermentation of tea extracts. Similarly, another study has successfully discriminated tea plant varieties based on their near- and visible infrared spectrum and using the ANN model [58].

NIR has been applied to obtain non-destructive measurements of TPC and TFC in various food commodities, such as coffee, honey, soybean, quinoa, etc. NIRS can be applied to estimate different parameters determining fruit quality, such as soluble chemical components, firmness, and acidity. The percentage of chlorogenic acid was determined in roasted coffee using HPLC–UV–NIR in the spectral range of 800–2500 nm along with RBF-NN chemometrics [59]. The phenolic content in honey was also determined by the NIR technique at 10 000–4000 cm^{-1} wavelength range with a high correlation coefficient value of 0.84 and maximum performance ratio of standard deviation, indicating the precision of NIR spectra for TPC estimation [60]. In addition, the soybean seed industry produces a by-product, soybean meal, commonly utilized in animal feeding materials or discarded directly in nature, and is responsible for creating environmental problems for authorized bodies (municipalities). Soybean meals were utilized as substrates for specific microorganisms to grow and synthesize biosurfactants, biodiesel, and many more, reducing the soybean seed industry. Moreover, soybean contains phenolic and flavonoid compounds that can be detected by vibrational spectroscopy. The process to detect TPC involves drying of soybean meal followed by grinding, sieving, maceration, and extraction of polyphenols. During this process, a microwave-assisted extraction technique was utilized to isolate polyphenols from soybean meals, providing excellent polyphenol yields and accessibility under sustainable environmental parameters [61]. TPC and TFC were assessed according to the process proposed by Dono and Indarto [62] and Padmajaaand Prasad [63], respectively, by using FTIR with MAE technique. Moreover, another study was conducted on the determination of TPC using UV–VIS–NIR spectroscopy [64].

12.4.2 MIR spectroscopy to determine polyphenols

MIR spectroscopy is also reported to have been used for determination of polyphenols in various food commodities. Total phenolic content, total anthocyanin content and ascorbic acid in pomegranate juice using UV-Vis MIR spectroscopy in the spectral range of 5000–930 nm. The results were then subjected to PLS model and a high correlation value, with a small room mean square error of calibration and predictions of 0.682, 0.13, and 0.15, respectively, were found [65]. Moreover, several anthocyanin concentrations of wine examined by MIR spectroscopy yielded reasonable prediction results (R^2 0.38–0.81), suggesting that this technology might be used to screen the changes of these compounds at various phases of the manufacturing process, particularly aging [66]. Synonymous study on wine was concluded on Australian Syrah wine to estimate total anthocyanins by building models [67]. Furthermore, using MIR spectroscopy, the microbiological and chemical parameters of wines can be predicted by employing PLS regression and filtering techniques, such as second-order derivation, thus yielding models with good correlation coefficients (R^2 > 0.86) and low RMSEP for total anthocyanin and phenolic contents [68].

12.4.3 FTIR spectroscopy to determine polyphenols

The conventional methods of polyphenols estimation in food requires chemical reagents and is a lengthy process [69]. IRS is nowadays used in the food and beverage industry as a fast and non-polluting analytical technique that uses no chemical reagents. The FTIR technique has also been employed for herbal and fruits/vegetables analysis. The polyphenolic content of six varieties of seagrasses in the Gulf of Mannar, India, has been measured [70]. In this study, the authors used FT MIRS to analyze gallic acid, p-coumaric acid, tannic acid, and vanillin in aqueous methanolic extract of seagrass. FTIR and UV–Vis spectroscopy has been used in the spectral range of 4000–400 cm^{-1}. In another study, total phenolic content in chickpea was determined using UV–FTIR in the spectral range of 4000–450 cm^{-1}[71]. In the case of okra seeds, TPC, Isoquercitrin, and Quercetin-3-O-gentiobiosides were determined using UV–VIS–FT-MIR spectroscopy [72]. A similar study was conducted to determine TPC, TFC, ascorbic acid content and TAC in apples, grape fruits, grape juice, pomegranate, and pomegranate juice with the help of FT-NIR and FT-MIR [73–76].

Another study was conducted on honey to determine catechins, chologenic acid, syringic acid, and vanillic acid using HPLC-DAD FTIR spectroscopy in the range 4000–600 cm^{-1} and the PLS model [12]. Meanwhile, total phenolic acid, epicatechin, and catechin were determined in chocolate using FT-IR and FT-MIR spectroscopy [77]. Similar studies were conducted on tea and coffee to detect their quality based on polyphenol content. Epigallocatechin gallate, epigallocatechin, epicatechin gallate, and epicatechin have been detected using NIR spectroscopy (4000–8500 cm^{-1}) and PLS [78]. Caffeic acid, catechin, cholorgenic acid and related compounds have been determined in coffee beans and spent coffee using HPLC-DAD and FTIR spectroscopy in conjugation with PLS statistical tool [79, 80]. In addition, olive oil samples were analyzed for hydroxytyrosol and derivatives, tyrosol and its derivatives, 3,4-DHPEA-EDA, o-Diphenols, p-HPEA-EDA, and total phenols using HPLC-DAD FT-NIR and PLS model [3].

FT-MIR spectroscopy in combination with PLS regressions has also been used to analyze total phenolic compounds, total anthocyanins, and total condensed tannins in wine samples throughout the red winemaking process [81]. Furthermore, because this approach does not need sample preparation or spectral pretreatment, it may be possible to provide an accurate assessment of the phenolic content of wines during the winemaking process, regardless of the stage of skin-contact maceration. In the case of wine samples, FT-MIR spectroscopy in conjunction with PLS regressions is reported to have been employed for assessment of total phenolic compounds, anthocyanins, and condensed tannins throughout the red winemaking [81]. These determinations were made in wine prepared from different grape varieties (Cabernet Sauvignon, Carignana, Merlot, Syrah, and Tempranillo) and harvested at different ripening stages, and the RMSEP and RPD values obtained demonstrated the usefulness of this technique for determining polyphenols. Furthermore, this technique allows detection without spectral pretreatment or sample preparation, thus it may feasibly offer a valid assessment of the phenolic content of wines during winemaking. Similarly, anthocyanins detection in wine have been determined using FT-MIR spectroscopy and PLS-R for rapid detection [82]. Synonymous studies have been conducted on red wine for detection on tannin using FT-MIR in the spectral range of 400–4000 cm^{-1} [83].

12.4.4 RAMAN spectroscopy to determine polyphenols

Apart from the other vibration spectroscopies, RS has also been used to determine food polyphenols. RS and FTIR have been used to study the secondary structural modification that arises from adding polyphenols and dietary fibers in bread products [84]. In addition, RS, together with NIRS, was presented as an online analysis tool for potato products and was used to directly determine total polyphenolic content, fat, color parameters, and antioxidant activity from colored-flesh potatoes [85]. Moreover, the polyphenolic content is high in medicinal plants, which are implicated in health benefits. *Potentilla tormentilla*, a therapeutic herb, is characterized by high tannins. The dried tormentil rhizomes are used for the production of herbal medicines. RS and mid- and near-IRS have been applied to quantify the active ingredients, such as total polyphenols, procyanidins, 3-O-galloylquininc acid, agrimoniin, and tannins, in *P. tormentilla* [86]. The same approach of using RS and other vibrational spectroscopy methods has been used to determine polyphenols and flavonoids in *Cistus incanus* leaves [87]. Wong *et al* have developed a rapid and non-destructive method for the quality control of *Pueraria lobata* and *Pueraria thomsonii* by determining the total polyphenolic content and starch composition of the two species using RS coupled with PLS analysis [88]. In addition, Espina *et al*, for the first time, provided the SERS spectra for four characteristic polyphenols of plant gallnuts; tannin, gallate, pyrogallol, and syringic acid [89]. Moreover, wine is rich in various polyphenols, which have been determined by RS. Polyphenols, tannins, and anthocyanins, present in Cabernet Sauvignon's variety of red wines, have been estimated using a portable Raman spectrometer. Raman spectra of different compounds of interest were obtained corresponding to different vineyards, harvest times, and seasons. A correlation of $R^2 = 0.895$ was obtained [90]. A synonymous study was conducted on honey and carrots to determine phenolic content catechins, chlorogenic acid, syringic acid and vanillic acid and carotenoids, α-Carotene, β-Carotene, lutein, polyacetylenes, falcarinol, falcarindiol, and 3-O-Acetylfalcarindiol using HPLC-DAD-RS in conjugation with PLS model [12, 91].

12.5 Future perspectives

Recent advances in vibrational spectroscopy have presented an opportunity for research groups, food industries, and pharmaceutical companies to form a well-established, non-destructive analytical, and eco-friendly methodology to examine functional bioactive and volatile compounds. IRS and RS have provided the technology for bioactive molecules assessment that requires no pretreatment of the analyte and does not use other environmentally harmful reagents. Although there are undeniable advantages of using vibrational spectroscopy as an analytical tool, some issues need to be considered if we aim for the laboratory-to-industrial translation of this spectroscopy technology. The spectral data obtained contains noises that need to be filtered out for proper interpretation of data and identification of the compound under investigation; therefore, there is a need for spectral data analysis procedures that can effectively remove this unusable information. In addition, interpreting spectral data requires a skilled and mathematically proficient operator, which is one of the limitations of the practical use of vibrational spectroscopy. Recent studies have used mathematical approaches such as PLS, RFA, and SPA in combination with IRS and RS to increase the efficiency of the technique in detecting and identifying bioactive molecules. However, if industrial use is desired, then further research is required to develop and optimize a mathematical prediction model for rapid, cheap, and online monitoring of industrial products. Although spectroscopic techniques are cheaper than conventional methods of analysis, the vibrational spectroscopic instrument is quite expensive. Ongoing studies aim to provide a more compact, cheaper, and more robust Raman and IR instrument for integration into the industrial process of product analysis during manufacturing and storage. Recent research has investigated the analytical potential of vibrational spectroscopy; nonetheless, further scientific research is expected to develop an inclusive range of new instruments to develop the application area.

12.6 Conclusion

This chapter has described the application and advantages of vibrational spectroscopy to estimate dietary polyphenol compounds in different food commodities. IR and RS have been used to investigate the chemical structures of sample components. Among the different vibrational techniques presented, NIRS has shown to be more practical in determining bioactive and volatile compounds than other spectroscopic techniques. Only a few studies have been conducted that have utilized RS for chemical species analysis. Vibrational spectroscopic techniques are much more straightforward, non-destructive, and cost-effective than traditional chemical methods for estimating the chemical components of the analyte. They can be used solo or in combination with chemometric tools for quantitative or qualitative analysis of samples, although the latter is preferred in studies because it provides an objective approach to compound identification. The main challenge restricting their use is that the spectral data of structurally similar substances are hard to distinguish. Recent studies have promoted vibrational spectroscopy as a versatile analytical technique to estimate different biochemical compounds present in foods, beverages, and plant materials, which determine their qualitative aspects such as organic acids, sugar, polyphenols, and vitamins. Vibrational spectroscopy has been used to examine bioactive molecules at the laboratory scale, in the qualitative characterization of raw/processed food attributes, and in online monitoring of products during the manufacturing process.

References

[1] Cozzolino D 2015 Infrared spectroscopy as a versatile analytical tool for the quantitative determination of antioxidants in agricultural products, foods and plants *Antioxidants* **4** 482–97

[2] Ayseli M T and Ayseli Y 2016 Flavors of the future: health benefits of flavor precursors and volatile compounds in plant foods *Trends Food Sci. Technol.* **48** 69–77

[3] Inarejos-García A M, Gómez-Alonso S, Fregapane G and Salvador M D 2013 Evaluation of minor components, sensory characteristics and quality of virgin olive oil by near infrared (NIR) spectroscopy *Food Res. Int.* **50** 250–8

[4] Baiz C R *et al* 2020 Vibrational spectroscopic map, vibrational spectroscopy, and intermolecular interaction *Chem. Rev.* **120** 7152–218

[5] Sathyanarayana D N 2015 *Vibrational Spectroscopy: Theory and Applications* (New Delhi: New Age International)

[6] Zhao X 2012 Advances and technology in infrared spectroscopy *J. Anqing Teachers College* **18** 94–7

[7] Skoog D A, Holler F J and Crouch S R 2017 *Principles of Instrumental Analysis* (Boston, MA: Cengage Learning)

[8] Sharma B, Frontiera R R, Henry A I, Ringe E and Van Duyne R P 2012 SERS: materials, applications, and the future *Mater. Today* **15** 16–25

[9] Zeisel D, Deckert V, Zenobi R and Vo-Dinh T 1998 Near-field surface-enhanced Raman spectroscopy of dye molecules adsorbed on silver island films *Chem. Phys. Lett.* **283** 381–5

[10] Kudelski A 2008 Analytical applications of Raman spectroscopy *Talanta* **76** 1–8

[11] Gao F, Xu L, Zhang Y, Yang Z, Han L and Liu X 2018 Analytical Raman spectroscopic study for discriminant analysis of different animal-derived feedstuff: understanding the high correlation between Raman spectroscopy and lipid characteristics *Food Chem.* **240** 989–96

[12] Tahir H E, Xiaobo Z, Zhihua L, Jiyong S, Zhai X, Wang S and Mariod A A 2017 Rapid prediction of phenolic compounds and antioxidant activity of Sudanese honey using Raman and Fourier transform infrared (FT-IR) spectroscopy *Food Chem.* **226** 202-–11

[13] Li-Chan E C, Ismail A A, Sedman J and Van de Voort F R 2002 Vibrational spectroscopy of food and food products *Handbook of Vibrational Spectroscopy* **vol 5** pp 3629–62

[14] Tahir H E, Xiaobo Z, Jianbo X, Mahunu G K, Jiyong S, Xu J L and Sun D W 2019 Recent progress in rapid analyses of vitamins, phenolic, and volatile compounds in foods using vibrational spectroscopy combined with chemometrics: a review *Food Anal. Methods* **12** 2361–82

[15] Schulz H and Baranska M 2007 Identification and quantification of valuable plant substances by IR and Raman spectroscopy *Vib. Spectrosc.* **43** 13–25

[16] Han X, Shen T and Lou H 2007 Dietary polyphenols and their biological significance *Int. J. Mol. Sci.* **8** 950–88

[17] Becker L B 2004 New concepts in reactive oxygen species and cardiovascular reperfusion physiology *Cardiovasc. Res.* **61** 461–70

[18] Gorinstein S, Zemser M, Weisz M, Halevy S, Deutsch J, Tilis K, Feintuch D, Guerra N, Fishman M and Bartnikowska E 1994 Fluorometric analysis of phenolics in persimmons *Bioscience, Biotechnology, and Biochemistry* **58** 1087–92

[19] Antolovich M, Prenzler P, Robards K and Ryan D 2000 Sample preparation in the determination of phenolic compounds in fruits *Analyst* **125** 989-–1009

[20] Butterfield D A, Castegna A, Pocernich C B, Drake J, Scapagnini G and Calabrese V 2002 Nutritional approaches to combat oxidative stress in Alzheimer's disease *J. Nutr. Biochem.* **13** 444–61

[21] Kim K H, Tsao R, Yang R and Cui S W 2006 Phenolic acid profiles and antioxidant activities of wheat bran extracts and the effect of hydrolysis conditions *Food Chem.* **95** 466–73

[22] Adom K K and Liu R H 2002 Antioxidant activity of grains *J. Agric. Food Chem.* **50** 6182–7

[23] Wang H J and Murphy P A 1994 Isoflavone content in commercial soybean foods *J. Agric. Food Chem.* **42** 1666–73

[24] Mazur W M, Duke J A, Wähälä K, Rasku S and Adlercreutz H 1998 Isoflavonoids and lignans in legumes: nutritional and health aspects in humans *J. Nutr. Biochem.* **9** 193–200

[25] Tsao R, Papadopoulos Y, Yang R, Young J C and McRae K 2006 Isoflavone profiles of red clovers and their distribution in different parts harvested at different growing stages *J. Agric. Food Chem.* **54** 5797–805

[26] Garazd M M, Garazd Y L and Khilya V P 2003 Neoflavones. 1. Natural distribution and spectral and biological properties *Chem. Nat. Compd.* **39** 54-–121

[27] Tsao R, Yang R, Young J C and Zhu H 2003 Polyphenolic profiles in eight apple cultivars using high-performance liquid chromatography (HPLC) *J. Agric. Food Chem.* **51** 6347–53

[28] Zhao F, Watanabe Y, Nozawa H, Daikonnya A, Kondo K and Kitanaka S 2005 Prenylflavonoids and phloroglucinol derivatives from hops (*Humulus lupulus*) *J. Nat. Prod.* **68** 43–9

[29] Laura A, Alvarez-Parrilla E and González-Aguilar G A 2009 *Fruit and Vegetable Phytochemicals: Chemistry, Nutritional Value and Stability* (New York: Wiley)

[30] Andersen O M and Markham K R (ed) *Flavonoids: Chemistry, Biochemistry and Applications* (Boca Raton, FL: CRC Press) 2005

[31] Williams C A 2006 Flavone and flavonol O-glycosides *Flavonoids: Chemistry, Biochemistry and Applications* pp 749–856

[32] Si W, Gong J, Tsao R, Kalab M, Yang R and Yin Y 2006 Bioassay-guided purification and identification of antimicrobial components in Chinese green tea extract *J. Chromatogr.* A **1125** 204-–10

[33] Prior R L, Lazarus S A, Cao G, Muccitelli H and Hammerstone J F 2001 Identification of procyanidins and anthocyanins in blueberries and cranberries (*Vaccinium* spp.) using high-performance liquid chromatography/mass spectrometry *J. Agric. Food Chem.* **49** 1270–6
[34] Davis C B, Markey C E, Busch M A and Busch K W 2007 Determination of capsaicinoids in habanero peppers by chemometric analysis of UV spectral data *J. Agric. Food Chem.* **55** 5925–33
[35] Bratt K, Sunnerheim K, Bryngelsson S, Fagerlund A, Engman L, Andersson R E and Dimberg L H 2003 Avenanthramides in oats (*Avena sativa* L.) and structure– antioxidant activity relationships *J. Agric. Food Chem.* **51** 594–600
[36] Koushki M, Amiri-Dashatan N, Ahmadi N, Abbaszadeh H A and Rezaei-Tavirani M 2018 Resveratrol: a miraculous natural compound for diseases treatment *Food Sci. Nutr.* **6** 2473–90
[37] Kang I, Buckner T, Shay N F, Gu L and Chung S 2016 Improvements in metabolic health with consumption of ellagic acid and subsequent conversion into urolithins: evidence and mechanisms *Adv. Nutr.* **7** 961–72
[38] Hewlings S J and Kalman D S 2017 Curcumin: a review of its effects on human health *Foods* **6** 92
[39] Wang S, Zhang C, Yang G and Yang Y 2014 Biological properties of 6-gingerol: a brief review *Nat. Prod. Commun.* **9** 1934578X1400900736
[40] Alfieri A and Mann G E 2015 Bioactive nutraceuticals and stroke: activation of endogenous antioxidant pathways and molecular mechanisms underlying neurovascular protection *Bioactive Nutraceuticals and Dietary Supplements in Neurological and Brain Disease* (New York: Academic) pp 365–79
[41] Cayuela J A and García J F 2017 Sorting olive oil based on alpha-tocopherol and total tocopherol content using near-infra-red spectroscopy (NIRS) analysis *J. Food Eng.* **202** 79–88
[42] Revilla I, Vivar-Quintana A M, González-Martín I, Escuredo O and Seijo C 2017 The potential of near infrared spectroscopy for determining the phenolic, antioxidant, color and bactericide characteristics of raw propolis *Microchem. J.* **134** 211–7
[43] Dong W, Ni Y and Kokot S 2013 A near-infrared reflectance spectroscopy method for direct analysis of several chemical components and properties of fruit, for example, Chinese hawthorn *J. Agric. Food Chem.* **61** 540–6
[44] Pissard A, Fernández Pierna J A, Baeten V, Sinnaeve G, Lognay G, Mouteau A, Dupont P, Rondia A and Lateur M 2013 Non-destructive measurement of vitamin C, total polyphenol and sugar content in apples using near-infrared spectroscopy *J. Sci. Food Agric.* **93** 238–44
[45] Ding X, Guo Y, Ni Y and Kokot S 2016 A novel NIR spectroscopic method for rapid analyses of lycopene, total acid, sugar, phenols and antioxidant activity in dehydrated tomato samples *Vib. Spectrosc.* **82** 1–9
[46] Caramês E T, Alamar P D, Poppi R J and Pallone J A 2017 Quality control of cashew apple and guava nectar by near infrared spectroscopy *J. Food Compos. Anal.* **56** 41–6
[47] Baca-Bocanegra B, Hernández-Hierro J M, Nogales-Bueno J and Heredia F J 2019 Feasibility study on the use of a portable micro near infrared spectroscopy device for the 'in vineyard' screening of extractable polyphenols in red grape skins *Talanta* **192** 353–9
[48] Martelo-Vidal M J and Vázquez M 2014 Determination of polyphenolic compounds of red wines by UV–VIS–NIR spectroscopy and chemometrics tools *Food Chem.* **158** 28–34

[49] Musial C, Kuban-Jankowska A and Gorska-Ponikowska M 2020 Beneficial properties of green tea catechins *Int. J. Mol. Sci.* **21** 1744
[50] Kumar P S, Basheer S, Ravi R and Thakur M S 2011 Comparative assessment of tea quality by various analytical and sensory methods with emphasis on tea polyphenols *J. Food Sci. Technol.* **48** 440–6
[51] Vuong Q V and Roach P D 2014 Caffeine in green tea: its removal and isolation *Sep. Purif. Rev.* **43** 155–74
[52] Bian M, Skidmore A K, Schlerf M, Wang T, Liu Y, Zeng R and Fei T 2013 Predicting foliar biochemistry of tea (*Camellia sinensis*) using reflectance spectra measured at powder, leaf and canopy levels *ISPRS J. Photogramm. Remote Sens.* **78** 148–56
[53] Luypaert J, Zhang M H and Massart D L 2003 Feasibility study for the use of near infrared spectroscopy in the qualitative and quantitative analysis of green tea, *Camellia sinensis* (L.) *Anal. Chim. Acta* **478** 303-–12
[54] Chen Q, Zhao J, Huang X, Zhang H and Liu M 2006 Simultaneous determination of total polyphenols and caffeine contents of green tea by near-infrared reflectance spectroscopy *Microchem. J.* **83** 42–7
[55] Hazarika A K, Chanda S, Sabhapondit S, Sanyal S, Tamuly P, Tasrin S, Sing D, Tudu B and Bandyopadhyay R 2018 Quality assessment of fresh tea leaves by estimating total polyphenols using near infrared spectroscopy *J. Food Sci. Technol.* **55** 4867–76
[56] Birenboim M, Chalupowicz D, Maurer D, Barel S, Chen Y, Falik E, Kengisbuch D and Shimshoni J A 2022 Optimization of sweet basil harvest time and cultivar characterization using near-infrared spectroscopy, liquid and gas chromatography, and chemometric statistical methods *J. Sci. Food Agric.* **102** 3325–35
[57] Kang W, Lin H, Jiang R, Yan Y, Ahmad W, Ouyang Q and Chen Q 2022 Emerging applications of nano-optical sensors combined with near-infrared spectroscopy for detecting tea extract fermentation aroma under ultrasound-assisted sonication *Ultrason. Sonochem.* **88** 106095
[58] Li X and He Y 2008 Discriminating varieties of tea plant based on Vis/NIR spectral characteristics and using artificial neural networks *Biosyst. Eng.* **99** 313–21
[59] Shan J, Suzuki T, Suhandy D, Ogawa Y and Kondo N 2014 Chlorogenic acid (CGA) determination in roasted coffee beans by near infrared (NIR) spectroscopy *Eng. Agric., Environ. Food* **7** 139–42
[60] Tahir H E, Xiaobo Z, Tinting S, Jiyong S and Mariod A A 2016 Near-infrared (NIR) spectroscopy for rapid measurement of antioxidant properties and discrimination of Sudanese honeys from different botanical origin *Food Anal. Methods* **9** 2631–41
[61] Woumbo C Y, Kuate D, Klang M J and Womeni H M 2021 Valorization of glycine max (soybean) seed waste: optimization of the microwave-assisted extraction (MAE) and characterization of polyphenols from soybean meal using response surface methodology (RSM) *J. Chem.* **2021** 1–2
[62] Dono N D and Indarto E Soy-milk waste with soybean meal dietary substitution: effects on growth performance and meat quality of broiler chickens
[63] Padmaja A and Prasad N B 2011 Pomegranate (*Punicagranatum* L.) peel extract as a source of natural antioxidant *J. Food Sci. Eng.* **3** 171
[64] Vickers N J 2017 Animal communication: when I'm calling you, will you answer too? *Curr. Biol.* **27** R713–5

[65] Arendse E, Fawole O A, Magwaza L S, Nieuwoudt H and Opara U L 2018 Comparing the analytical performance of near and mid infrared spectrometers for evaluating pomegranate juice quality *LWT* **91** 180–90
[66] Sen I, Ozturk B, Tokatli F and Ozen B 2016 Combination of visible and mid-infrared spectra for the prediction of chemical parameters of wines *Talanta* **161** 130–7
[67] Ristic R, Cozzolino D, Jeffery D W, Gambetta J M and Bastian S E 2016 Prediction of phenolic composition of shiraz wines using attenuated total reflectance mid-infrared (ATR-MIR) spectroscopy *Am. J. Enol. Viticult.* **67** 460–5
[68] Canal C and Ozen B 2017 Monitoring of wine process and prediction of its parameters with mid-infrared spectroscopy *J. Food Process Eng* **40** e12280
[69] Zhang Q, Zhang J, Shen J, Silva A, Dennis D A and Barrow C J 2006 A simple 96-well microplate method for estimation of total polyphenol content in seaweeds *J. Appl. Phycol.* **18** 445–50
[70] Ragupathi Raja Kannan R, Arumugam R and Anantharaman P 2011 Fourier transform infrared spectroscopy analysis of seagrass polyphenols *Curr. Bioact. Compd.* **7** 118–25
[71] Kadiroğlu P, Aydemir L Y and Akcakaya F G 2018 Prediction of functional properties of registered chickpea samples using FT-IR spectroscopy and chemometrics *Lwt* **93** 463–9
[72] Xia F, Li C, Zhao N, Li H, Chang Q, Liu X, Liao Y and Pan R 2018 Rapid determination of active compounds and antioxidant activity of okra seeds using Fourier transform near infrared (FT-NIR) spectroscopy *Molecules* **23** 550
[73] Giovanelli G, Sinelli N, Beghi R, Guidetti R and Casiraghi E 2014 NIR spectroscopy for the optimization of postharvest apple management *Postharvest Biol. Technol.* **87** 13–20
[74] Li Y, You X and Shi X 2017 Enhanced chemiluminescence determination of hydrogen peroxide in milk sample using metal–organic framework Fe–MIL–88NH 2 as peroxidase mimetic *Food Anal. Methods* **10** 626–33
[75] Caramês E T, Alamar P D, Poppi R J and Pallone J A 2017 Rapid assessment of total phenolic and anthocyanin contents in grape juice using infrared spectroscopy and multivariate calibration *Food Anal. Methods* **10** 1609-–15
[76] Arendse E, Fawole O A, Magwaza L S, Nieuwoudt H H and Opara U L 2017 Development of calibration models for the evaluation of pomegranate aril quality by Fourier-transform near infrared spectroscopy combined with chemometrics *Biosystems Eng.* **159** 22–32
[77] Hu Y, Pan Z J, Liao W, Li J, Gruget P, Kitts D D and Lu X 2016 Determination of antioxidant capacity and phenolic content of chocolate by attenuated total reflectance-Fourier transformed-infrared spectroscopy *Food Chem.* **202** 254–61
[78] Chen Q, Zhao J, Chaitep S and Guo Z 2009 Simultaneous analysis of main catechins contents in green tea (*Camellia sinensis* (L.)) by Fourier transform near infrared reflectance (FT-NIR) spectroscopy *Food Chem.* **113** 1272–7
[79] Magalhães L M, Machado S, Segundo M A, Lopes J A and Páscoa R N 2016 Rapid assessment of bioactive phenolics and methylxanthines in spent coffee grounds by FT-NIR spectroscopy *Talanta* **147** 460–7
[80] Liang N, Lu X, Hu Y and Kitts D D 2016 Application of attenuated total reflectance–Fourier transformed infrared (ATR-FTIR) spectroscopy to determine the chlorogenic acid isomer profile and antioxidant capacity of coffee beans *J. Agric. Food Chem.* **64** 681–9
[81] Fragoso S, Acena L, Guasch J, Mestres M and Busto O 2011 Quantification of phenolic compounds during red winemaking using FT-MIR spectroscopy and PLS-regression *J. Agric. Food Chem.* **59** 10795–802

[82] Romera-Fernández M, Berrueta L A, Garmón-Lobato S, Gallo B, Vicente F and Moreda J M 2012 Feasibility study of FT-MIR spectroscopy and PLS-R for the fast determination of anthocyanins in wine *Talanta* **88** 303––10
[83] Fernández K and Agosin E 2007 Quantitative analysis of red wine tannins using Fourier-transform mid-infrared spectrometry *J. Agric. Food Chem.* **55** 7294–300
[84] Sivam A S, Sun-Waterhouse D, Perera C O and Waterhouse G I 2013 Application of FT-IR and Raman spectroscopy for the study of biopolymers in breads fortified with fibre and polyphenols *Food Res. Int.* **50** 574–85
[85] Mazurek S, Szostak R, Kita A, Kucharska A Z, Sokół-Łętowska A and Hamouz K 2017 Determination of antioxidant activity and polyphenols content in chips by Raman and IR spectroscopy *Food Anal. Methods* **10** 3964–71
[86] Mazurek S, Fecka I, Węglińska M and Szostak R 2018 Quantification of active ingredients in *Potentilla tormentilla* by Raman and infrared spectroscopy *Talanta* **189** 308––14
[87] Pielorz S, Fecka I, Bernacka K and Mazurek S 2022 Quantitative determination of polyphenols and flavonoids in Cistus × incanus on the basis of IR, NIR and Raman spectra *Molecules* **28** 161
[88] Wong K H, Razmovski-Naumovski V, Li K M, Li G Q and Chan K 2015 The quality control of two Pueraria species using Raman spectroscopy coupled with partial least squares analysis *J. Raman Spectrosc.* **46** 361–8
[89] Espina A, Sanchez-Cortes S and Jurašeková Z 2022 Vibrational study (Raman, SERS, and IR) of plant gallnut polyphenols related to the fabrication of iron gall inks *Molecules* **27** 279
[90] Gallego Á L, Guesalaga A R, Bordeu E and González Á S 2010 Rapid measurement of phenolics compounds in red wine using Raman spectroscopy *IEEE Trans. Instrum. Meas.* **60** 507––12
[91] Krähmer A, Böttcher C, Rode A, Nothnagel T and Schulz H 2016 Quantifying biochemical quality parameters in carrots (*Daucus carota* L.)–FT-Raman spectroscopy as efficient tool for rapid metabolite profiling *Food Chem.* **212** 495–502

www.ingramcontent.com/pod-product-compliance
Ingram Content Group UK Ltd.
Pitfield, Milton Keynes, MK11 3LW, UK
UKHW051857100926
8556IPUK00005B/669